2012 年上海市文化创意产业扶持项目

STYLE SHANGHAI

2015春夏海派时尚流行趋势
2015 SPRING/SUMMER STYLE SHANGHAI FASHION TREND

服装篇
APPAREL

海派时尚流行趋势研究中心　著

东华大学出版社

[服装篇]

编辑委员会
名誉主任： 厉无畏
顾问： 徐明稚、Christopher Breward、席时平
主任： 刘春红
执行主任： 刘晓刚
副主任： 刘健、贺寿昌、杲云、朱勇
委员： 林艺、庄培、吕苏宁、邵峰、蒋智威、徐晶
　　　　卞向阳、吴翔、彭波、汪芳、钟宏、吉承
统筹： 李峻、曹霄洁、吴亮、傅白璐
编排： 刘清芸、胡弘毅、周吉、范乃文

主创人员：
[男装]
罗竞杰、厉莉、张颖、陈彦静婷、郭云昕、刘芮希、钱昳婷、余苑、俞梦婷
[女装]
蔡凌霄、刘灿明、边菲、商斯云、陶媛沁、崔彦、曾芷茜、徐方昕、刘勤艺、高宁馨、毛江天、吴滨彬

支持单位：
上海市经济和信息化委员会
上海市文化创意产业推进领导小组办公室
上海市长宁区人民政府
东华大学
上海智富企业发展（集团）有限公司
司色艾印染科技（上海）有限公司
上海国际服装服饰中心

翻译：
Translation Graciously Provided by:The Odyssey Network
Steve Zades （Donghua University Consultant Professor for Postgraduates,USA）
Zheng Yuan Mao

目录 CONTENTS

序 PREFACE	4
2015 春夏海派时尚流行趋势主题：梦 2015 SPRING/SUMMER STYLE SHANGHAI FASHION TREND THEMES：DREAMS	6
男装 MENSWEAR	8
海上星梦 STAR DREAMS OVER THE SEA	10
优活之梦 LIVING THE DREAM	28
时空筑梦 REALISING DREAMS THROUGH SPACE	46
多彩享梦 ENJOY COLORFUL DREAMS	64
女装 WOMENSWEAR	82
海上星梦 STAR DREAMS OVER THE SEA	84
优活之梦 LIVING THE DREAM	102
时空筑梦 REALISING DREAMS THROUGH SPACE	120
多彩享梦 ENJOY COLORFUL DREAMS	138
2015 春夏海派时尚流行要素汇编·服装篇 2015 SPRING/SUMMER STYLE SHANGHAI FASHION ELEMENTS COLLECTION·APPAREL	156

序

流行趋势植根于时代，时尚潮流溯源于文化，自"海派时尚"创刊号成功发布后，创作团队就海派时尚的文化与时代的交汇进行了追根溯源和深入探讨。透过海派电影、海派文学、海派音乐、海派戏剧……发现海派时尚的文化之源。

电影，是时尚文化重要的组成部分，是塑造精神之梦的艺术体裁，更是时尚设计师们热衷的灵感源泉。海派电影，在中国电影发展历程中有着不可替代的重要地位。自1897年于上海天华茶园电影首次传入中国，20世纪30年代风华绝伦的电影明星胡蝶、阮玲玉等的出现，到今天海纳百川的电影新纪元，都成为设计师们枝繁叶茂的灵感花园。2015年春夏海派时尚流行趋势以海派电影为主线，分别从海派经典风格、海派自然风格、海派未来风格、海派都市风格等方面进行演绎。

为了方便时尚业者按图索骥，分类检索，2015年春夏海派时尚流行趋势分为三大篇章，分别为灵感篇、服装篇和配饰篇。其中，灵感篇包括海派文化、面料、图形章节，提供海派时尚趋势设计的灵感和素材；服装篇包括海派男装和女装章节；配饰篇包括海派鞋履、箱包、帽饰、首饰等章节。

作为上海市文化创意产业扶植项目、环东华时尚创意产业服务平台的重要内容，本书的出版得到了上海市经信委、长宁区政府、东华大学各级领导和社会各界的广泛支持。团队以东华大学设计学科专家为主体，汇集了艺术家、设计师、企业家、国际友人，团队成员近百名，共同迸发出创意火花，循序渐进，数易其稿，历时半年完成。

时尚，是当代人所崇尚的一种态度、一种文化、一种精神，更是一种生活方式。时尚中国梦，是中国人拥有自己的原创风格、流行趋势、创意设计，在世界时尚界有属于自己的时尚话语权。坚持原创驱动，打造海派时尚，建设美丽中国！

海派时尚流行趋势研究中心

Preface

While changing with times, fashion trends have always been strongly influenced by social and cultural movements. Since successfully launching the first issue of "Style Shanghai", our team have continued to trace and further study the origins of Shanghai style from different aspects including movies, literature, music, drama and etc.

Film, an important part of fashion culture, is not only what we use to make dream come true in virtual world, but also a popular resource of design inspiration. Shanghai Style movies hold an irreplaceable position in the Chinese film history. Since movie was first publicly introduced to Chinese people in Shanghai Tianhua Tea House in 1897, this industry has developed a lot from unforgettable 1930s movie stars, such as Die Hu and Lingyu Ruan, famous for their incredible beauty and talent to thousands of films of different genres, which have inspired our designers. This book will tell you many design stories behind our work with Shanghai Style movie as a main theme and other 4 subthemes, classical style, natural style, futuristic style and urban style.

For your convenience, we have tailored this book with 3 chapters, Inspiration, Apparel and Accessories. Inspiration includes articles on Shanghai style culture, fabric and graphics; Apparel is divided into men's and women's; Accessories covers footwear, bag, hat, jewelry and etc.

As a project to foster Shanghai culture and innovation industries and a main product of Greater Donghua service platform for innovation industry development, this book's publication has received a lot of support from Shanghai Municipal Commission of Economy and Information, Bureau of Changning District, Donghua University and other sectors of society. Our design team is mainly composed of experts from Donghua University, bringing together artists, designers, entrepreneurs and international friends. With almost 100 contributors' 6-month collaboration and continuous efforts, we represent you this second issue of Style Shanghai Fashion Trend.

Fashion tells about attitude, culture, spirit and way of living. Chinese fashion dreams are about having our own styles, trends, creative designs and fashion identity on world stage. In order to realize those dreams, we will persist in originality, inherit the essence from the tradition, popularize Style Shanghai fashion and build a more beautiful China!

Style Shanghai Fashion Trend Research Center

2015春夏海派时尚流行趋势主题
2015 SPRING/SUMMER STYLE SHANGHAI FASHION TREND THEMES

梦 DREAMS

电影是造梦的艺术，在国产电影走向世界的进程中，梦想的轮廓也愈加圆满。作为海派时尚独特表达的海派电影，是海派时尚设计的重要资源。在海派电影摇曳的光影中，海派时尚的脉络也愈发清晰。

Film is the art of dreaming, in the process of Chinese films stepping to the world, the outline of dream is becoming even more complete. As a unique fashion expression, Shanghai movie is an important resource for Style Shanghai fashion design. Shanghai movie swaying in light and shadow, the context of Style Shanghai is also increasingly clear.

海上星梦 [海派经典风格] STAR DREAMS OVER THE SEA [SHANGHAI CLASSICAL STYLE]

近代海派电影的繁荣造就了摩登生活方式的汇集，创新设计与怀旧经典相互交融。光影流动的时尚教科书蛊惑着沪上各阶层女性，她们很快成为电影所传诉的风尚追逐者和推波助澜者。

Modern Shanghai Movies created a modern lifestyle collection, innovative design and retro classics mingled. Light flowing fashion textbook charms the Shanghai women, and they soon became the movie fashion chaser.

优活之梦 [海派自然风格] LIVING THE DREAM [SHANGHAI NATURAL STYLE]

弄堂生活是水泥森林中的桃花源。在申城日渐国际化的今天，反映海派市井日常生活的影视剧持续风靡，也反映了当代人对慢活、乐活、优活理念的秉持。来自家庭手工式、生态环保的弄堂作品是现代都市人心中最温暖的城市记忆。

Alley life is paradise in concrete jungle. While Shanghai is becoming increasingly international, the movies of Shanghai daily life continues to be popular, reflecting contemporary people's persistence for slow living, lohas, and quality of life. Alley inspired handmade style from family, eco-friendly design work is the warmest memories of the modern city.

时空筑梦 [海派未来风格] REALISING DREAMS THROUGH SPACE [SHANGHAI FUTURISTIC STYLE]

高科技从思维、行为到审美的各个方面对人们的衣食住行产生影响。炫目的科幻大片以强烈的视觉冲击力牢牢吸引着城市年轻群体的目光。强规则几何图形、固化的符号、机械生物等是人们对未来生活的深思与肆意畅想。

High-tech is having impact on people's daily life, from thinking, behavior to the aesthetic aspects. Stunning sci-fi blockbuster with strong visual impact firmly attracted the attention of young urbanites. Strong geometry, curing symbols, mechanical and biological life are composing thinkings about the future.

多彩享梦 [海派都市风格] ENJOY COLORFUL DREAMS [SHANGHAI URBAN STYLE]

坚持个人理想、追求幸福与欢乐的个人价值的都市人，用个性打破思维的界限，在叛逆和玩乐中创意。多彩的感官享受，幽默、趣味和诙谐都让我们的"快乐崇拜"不停歇。

Adhere to personal dream, with pursuit of happiness and joy, individual thinkings to break the boundaries of creativity and play in the rebellion. Colorful sensuality, humor, fun and witty, makes our "Happy worship" never stop.

2015 SPRING/SUMMER STYLE SHANGHAI FASHION TREND

经典
CLASSICAL

海上星梦
STAR DREAMS OVER THE SEA
海派经典风格
SHANGHAI CLASSICAL STYLE

优活之梦
LIVING THE DREAM
海派自然风格
SHANGHAI NATURAL STYLE

西方　　　　　　　　　　　　　　　　　　　　　　　　　　　　东方
WEST　　　　　　　　　　　　　　　　　　　　　　　　　　　　EAST

时空筑梦
REALISING DREAMS THROUGH SPACE
海派未来风格
SHANGHAI FUTURISTIC STYLE

多彩享梦
ENJOY COLORFUL DREAMS
海派都市风格
SHANGHAI URBAN STYLE

未来
FUTURE

男装 Hi 上海设计之都公共服务平台 认定专业平台 |

MANSWEAR

海上星梦
STAR DREAMS OVER THE SEA
优活之梦
LIVING THE DREAM
时空筑梦
REALISING DREAMS THROUGH SPACE
多彩享梦
ENJOY COLORFUL DREAMS

海上星梦
STAR DREAMS OVER THE SEA

20世纪30年代的上海是淘金者们的乐土，冒险家的乐园。时代变迁带来都市新篇章，经典与摩登的融合，为经典塑造新的视觉飨宴，让我们一起来重温那个繁华时代。

In the 1930s, the old Shanghai is paradise for fortune seekers and adventurers. Changing times have brought a new chapter to the city, when the classic met the modern in design, a mixed style shaped. Let's take a look at the bustling era through some designs well integrating elements from different times.

MENSWEAR
2015 SPRING/SUMMER STYLE SHANGHAI FASHION TREND
海派男装流行趋势

2015 SPRING/SUMMER STYLE SHANGHAI FASHION TREND / MENSWEAR

关键要素 | KEY POINTS

繁华时代
FLOURISHING ERA

精细品质
EXQUISITE QUALITY

新贵气质
NOBLE TEMPERAMENT

创新传统
INNOVATING TRADITIONS

珍珠白

雷雨灰

麦糖黄

恒星橙

紫砂棕

紫檀棕

枣泥红

曜石黑

青砖灰

铁艺灰

S 男装 2015 春夏海派时尚流行趋势

海上星梦
STAR DREAMS OVER THE SEA

外滩万国建筑群见证上海滩的繁荣岁月

怀旧经典细格纹套装
黑色经典大衣
纯色高支棉衬衫

跑马场的看客满心期待胜利者的归来

奢华绚丽的百乐门是老上海纸醉金迷生活的缩影

《申报》记录了上海

设计灵感

20世纪30年代上海滩歌舞升平、纸醉金迷,而如今金融大亨、老克勒的身影在思绪中变的模糊不清,复古风貌的元素让人们重温那个繁华时代。

1930s, socialites lived in luxury and enjoyed a peachy life. Now in our memory, they are walking away under the dim stage light and only vintage can bring us some traces of that flouring era.

12 www.style.sh.cn

2015 SPRING/SUMMER STYLE SHANGHAI FASHION TREND / MENSWEAR

关键要素 | 繁华时代
KEY POINTS | FLOURISHING ERA

爵士乐奏响了黄金年代的序曲

双排四粒扣戗驳领西服
异色领衬衫

精致毛料西服展现尊贵的绅士气质

缎面青果领精致合体礼服
胸前塔克装饰礼服衬衫

沙砾黄
杏仁黄
亚铜灰
青砖灰
紫砂棕
紫檀棕
枣泥红
曜石黑

做旧风格方砖

S 男装

2015 春夏海派时尚流行趋势

海上星梦
STAR DREAMS OVER THE SEA

异色领条纹衬衫

撞色条纹马甲

一粒扣经典格纹青果领单西

菱形格长袖针织衫

迷你格纹卷口裤

2015 SPRING/SUMMER STYLE SHANGHAI FASHION TREND / MENSWEAR

关键要素 | 繁华时代
KEY POINTS | FLOURISHING ERA

宽格纹戗驳领西服
格纹印花衬衫
直筒西裤

一粒扣戗驳领单西
白色直筒裤
礼服衬衫

海上星梦
STAR DREAMS OVER THE SEA

高档精致的商务会所

高支温莎领衬衫
高驳口窄驳领一粒扣西套装

打造金融家风貌的精细品质

拔地而起的高楼大厦是精工技艺的展现

能遥望浦江景色的精致

设计灵感

淘金者们在现代化摩天大楼、老上海洋房中享受着属于自己的精致生活，得体的装束、精细的品质营造出属于这个时代的金融家风貌。

New-age pursuers of success enjoy exquisite life in modern skyscrapers and old Shanghai villas. They wear decent clothing and always go after fine qualities. This is exactly what modern bankers look like.

2015 SPRING/SUMMER STYLE SHANGHAI FASHION TREND / MENSWEAR

关键要素 | 精细品质
KEY POINTS | EXQUISITE QUALITY

精密仪器展现高超技艺

拼色戗驳领长大衣
V型领薄型针织衫
纯白色衬衫

青色细条纹
纯驳领西套装
纯棉纯色衬衫

摩天大楼展现高超建造技艺

浪花白
剑麻灰
雷雨灰
水晶蓝
青瓷灰
铁艺灰
陨石棕
瓷瓦黑

海上星梦
STAR DREAMS OVER THE SEA

短袖几何提花针织商务POLO衫

高支全棉条纹商务衬衫

平驳领两粒扣条纹商务单西

异色领异色克夫条纹商务衬衫

薄型V型领开襟针织衫

2015 SPRING/SUMMER STYLE SHANGHAI FASHION TREND / MENSWEAR

关键要素 | 精心品质
KEY POINTS | EXQUISITE QUALITY

经典格纹三件套
宽领面戗驳领
经典条纹衬衫

2015 春夏海派时尚流行趋势

海上星梦
STAR DREAMS OVER THE SEA

新古典主义家居设计

满地印花纹样西套装
深色圆领针织衫
印花衬衫

马勒别墅展现欧式建筑古典风貌

雪茄文化体现新贵

20世纪30年代古董家具营造出小资生活

设计灵感

当新古典主义遇到海派文化,迸发出新贵族气息。繁复、高贵的纹样,含蓄、典雅的色彩,打造出都市新贵形象。

When neoclassic style meets Shanghai style, they create a new burst of neo-nobility. The image of new urban upstart is composed of complex and elegant patterns, subtle and graceful colors.

海上星梦
STAR DREAMS OVER THE SEA

S 男装 2015 春夏海派时尚流行趋势

海上星梦
STAR DREAMS OVER THE SEA

时代变迁带来都市新篇章

兼具经典与摩登的别具一格

复古风貌西装外套
低饱和度恤衫
麻质宽松休闲裤

传统元素遇到现代设计带来的新风貌

设计灵感

时代变迁带来都市新篇章，传统技艺创新带来新风貌，为经典塑造新的视觉形象，用创新思维向经典致敬。

Changes in times have opened up a new chapter for the city. Innovations in traditional techniques start to spring up. Today, we are going to pay our homage to traditions innovatively.

2015 SPRING/SUMMER STYLE SHANGHAI FASHION TREND / MENSWEAR

关键要素 | 创新传统
KEY POINTS | INNOVATING TRADITIONS

传统的创新运用与新的表现形式给设计带来活力

新技术为针织产品带来新形象

经典格纹
全身服饰采用
统一经典纹样
的现代运用

多媒体感官体验

绞花纹样西外套
圆领恤衫
直筒西裤

- 豆浆白
- 香芋灰
- 铁艺灰
- 淡奶黄
- 斗笠黄
- 苔藓绿
- 抹茶绿
- 乌梅黑

S男装 2015 春夏海派时尚流行趋势

海上星梦
STAR DREAMS OVER THE SEA

假两件套长袖衬衫

梭织衬衫领长袖针织衫

西装风貌夹克

假两件长袖V领衬衫

假两件长款外套

假领带长袖衬衫

2015 SPRING/SUMMER STYLE SHANGHAI FASHION TREND / MENSWEAR

关键要素 | 创新传统
KEY POINTS | INNOVATING TRADITIONS

粗棒针开襟毛衫

经典双排扣戗驳领针织衫
圆领T恤
直筒休闲裤

针织西外套
V领针织开衫
复古纹样印花衬衫

S男装 2015 春夏海派时尚流行趋势

优活之梦
LIVING THE DREAM

上海的小弄堂里生活着许多平凡的上海人们，虽然生活空间狭小，收入并不富裕，但是他们却生活得井井有条，不经意间流露出随性悠闲的生活态度。

Many citizens are still living in Shanghai alleys trying to keep everything in perfect order, inadvertently revealing their attitude towards life despite of limited living space and below-average income.

MENSWEAR

2015 SPRING/SUMMER STYLE SHANGHAI FASHION TREND
海派男装流行趋势

S 男装 2015 春夏海派时尚流行趋势

优活之梦
LIVING THE DREAM

木质手缝纽扣

异质拼接设计
异质贴袋格纹夹克
小格纹衬衫

织补旧衣是中国的传统美德

上海里弄居民具有合理而节俭的生活习惯

上海街头常见的生活场景

设计灵感

节俭是中华民族的美德，延续 20 世纪七八十年代风靡上海的假领、袖套等再利用设计，有意识地做旧拼接、补丁工艺不经意间流露出随性悠闲的生活态度。

Frugality is a virtue of the traditional Chinese family. The re-use design of fake collar or sleeves, which were very popular in Shanghai during 70s and 80s, is the continuation of that virtue. Distressed stitching and patches inadvertently reveal sort of casual and carefree attitude towards life.

优活之梦
LIVING THE DREAM

2015 SPRING/SUMMER STYLE SHANGHAI FASHION TREND / MENSWEAR

关键要素 | 织补风尚
KEY POINTS | BACK TO THE BASICS

S男装 2015 春夏海派时尚流行趋势

优活之梦
LIVING THE DREAM

上海老城区中的休闲生活场景

有氧运动，放松心情的优质生活方式

柔软材质
异质拼接格纹单西
五分休闲裤

午后弄堂里梧桐树荫下平凡的市民生活场景

设计灵感

弄堂里、"下只角"生活着许许多多平凡的上海人，虽然他们的生活空间逼仄，但是起居生活并不怠慢，在随意之间也能流露出不一样的气质。

Many Shanghainese are still leading an alley-life in areas that haven't been reconstructed. Although their living space is cramped, their attitude toward elegant daily life is not bound. If you observe their ways of living carefully, sometimes you will smile at their wisdom and be pleased by their temperament, different from what is owned by people of noble birth.

优活之梦
LIVING THE DREAM

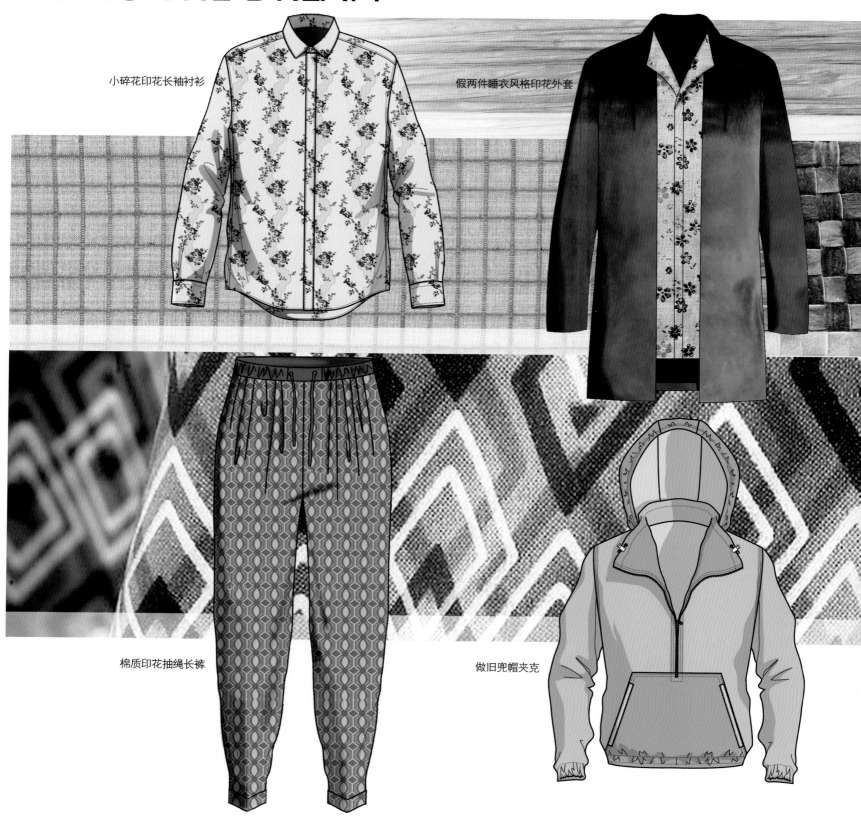

小碎花印花长袖衬衫

假两件睡衣风格印花外套

棉质印花抽绳长裤

做旧兜帽夹克

优活之梦
LIVING THE DREAM

大自然葱郁的景色

取自自然灵感与材质的金字塔造型模具

天然苎麻编织成的网状手提袋

森林木质肌理

全棉细条纹兜帽外套
纯白色圆领恤衫
棉质休闲短裤

设计灵感

光影绰绰的梧桐绿荫、波光粼粼的湖水，大自然带给人们的视觉体验让我们深刻反思生活的意义，质朴的色彩、自然的材质、宽松的廓型带来生态体验。

Nature gives us such fascinating visual experience and lets us reconsider about the meaning of life full of steel and cement. Plain colors, natural materials, and less polished shapes all bring us eco-experience.

2015 SPRING/SUMMER STYLE SHANGHAI FASHION TREND / MENSWEAR

关键要素 KEY POINTS | 生态材质 ECO MATERIAL

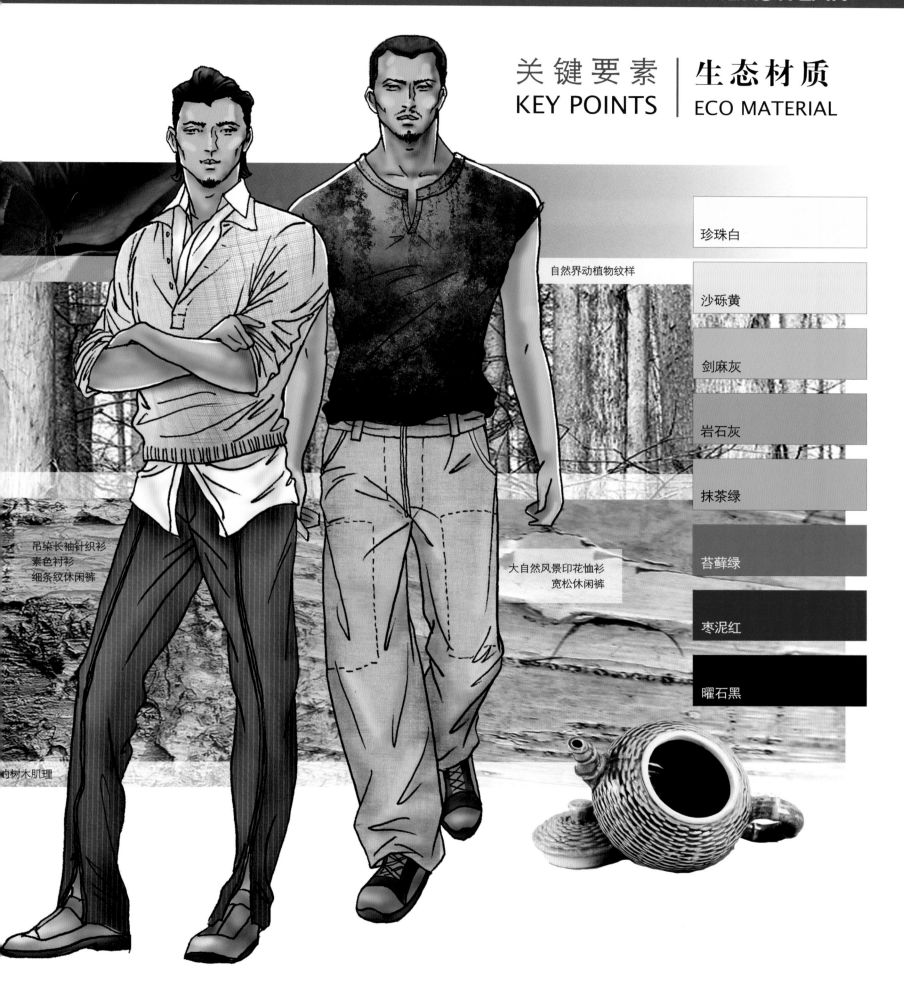

自然界动植物纹样

吊染长袖针织衫
素色衬衫
细条纹休闲裤

大自然风景印花恤衫
宽松休闲裤

约树木肌理

珍珠白
沙砾黄
剑麻灰
岩石灰
抹茶绿
苔藓绿
枣泥红
曜石黑

优活之梦
LIVING THE DREAM

插肩袖麻质补丁衬衫

薄型做皱驳领西装外套

多口袋立领做皱夹克

薄型插肩袖卷边针织衫

2015 SPRING/SUMMER STYLE SHANGHAI FASHION TREND / MENSWEAR

关键要素 | 生态材质
KEY POINTS | ECO MATERIAL

多口袋柔软夹克
V型领T恤
直筒休闲裤

做旧风格夹克
自然印花T恤
宽松多口袋休闲裤

S男装 2015 春夏海派时尚流行趋势

优活之梦
LIVING THE DREAM

手工打造的置物箱

精美绝伦的手工技艺

自然材质通过手工加工带来新的家具风格

拼接设计
格纹拼接马甲
格纹衬衫
做旧牛仔裤

设计灵感

刺绣、拼布绣、扎染等精湛手工艺在服装中的应用，温馨的色彩、轻松的图案，让穿着者回忆起对母亲、家庭的思念。

Application of exquisite handicrafts such as embroidery, quilting embroidery and tie-dye, together with warm colors and simple patterns, helps recall the sweets as being loved by mother and other families and also the bitters in the taste of missing.

优活之梦
LIVING THE DREAM

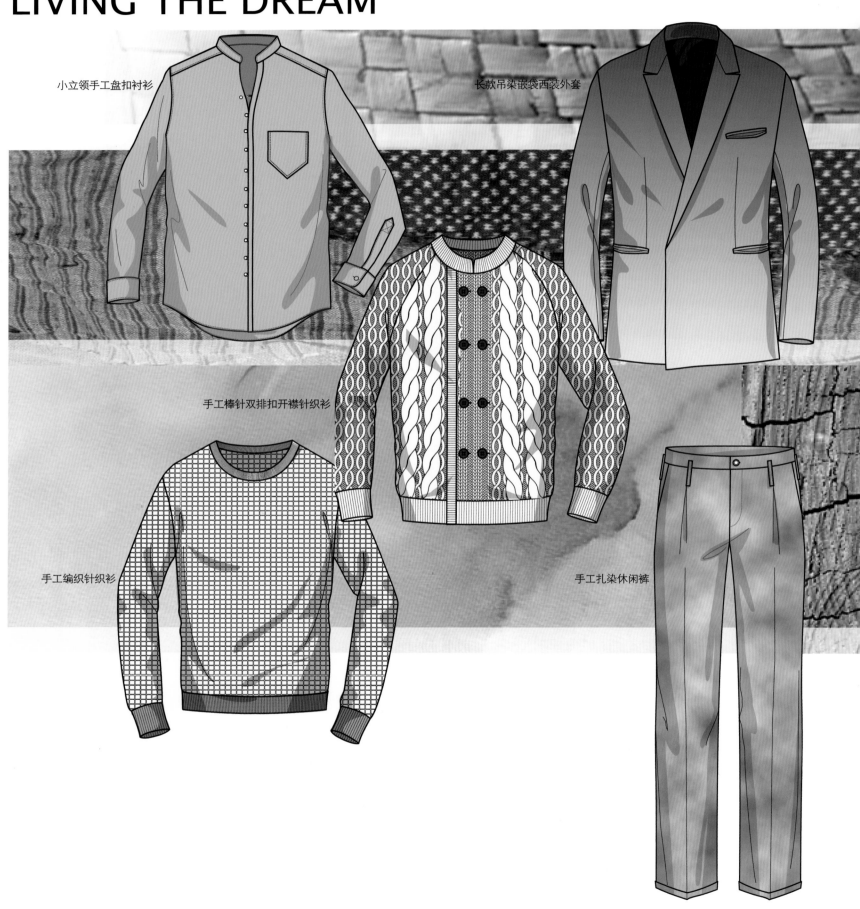

小立领手工盘扣衬衫

长款吊染嵌袋西装外套

手工棒针双排扣开襟针织衫

手工编织针织衫

手工扎染休闲裤

2015 SPRING/SUMMER STYLE SHANGHAI FASHION TREND / MENSWEAR

关 键 要 素 | 家庭手工
KEY POINTS | FAMILY HANDICRAFTS

数码印花衬衫

双面长款风衣
扎染恤衫
纯色棉质衬衫
格纹卷口休闲短裤

缉线装饰薄型针织衫
休闲短裤

S 男装 2015 春夏海派时尚流行趋势

时空筑梦
REALISING DREAMS THROUGH SPACE

单一元素的纯粹应用，各种基本元素的秩序重复，仿生学的应用，网络时代的变迁，借鉴造物者的杰作创造出超凡的作品。

Application of a single element, repetition of various basic elements, bionics, changes brought by information age all can be inspirations for an extraordinary work design.

MENSWEAR

2015 SPRING/SUMMER STYLE SHANGHAI FASHION TREND
海派男装流行趋势

2015 SPRING/SUMMER STYLE SHANGHAI FASHION TREND / MENSWEAR

关键要素 | KEY POINTS

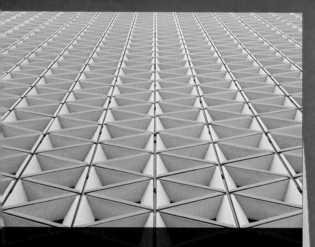

复制粘贴
REPETITION & MODULIZATION

符号人生
SYMBOLIC LIFE

机械生物
MECHANICAL CREATURE

单纯设计
SIMPLE DESIGN

珍珠白

氯化灰

岩石灰

苹果绿

柠檬黄

碳钢灰

天光蓝

光影蓝

恒星橙

陨石棕

时空筑梦
REALISING DREAMS THROUGH SPACE

重复三角造型平面构成感建筑物穹顶

几何形立面错落布局营造未来风格建筑物

光影构成的几何纹样打造构成主义

波尔卡圆点满地纹样
黑白色彩
H型廓型

设计灵感

单一、简单的几何元素不断地、有规则地重复着，就像细胞分裂一样复制扩张，组成的矩阵图形带来视觉上的冲击体验。

Repeating a simple geometric element constantly and regularly, just the same as the cell dividing and expanding, will make us matrix patterns and bring our eyes visual impacts.

时空筑梦
REALISING DREAMS THROUGH SPACE

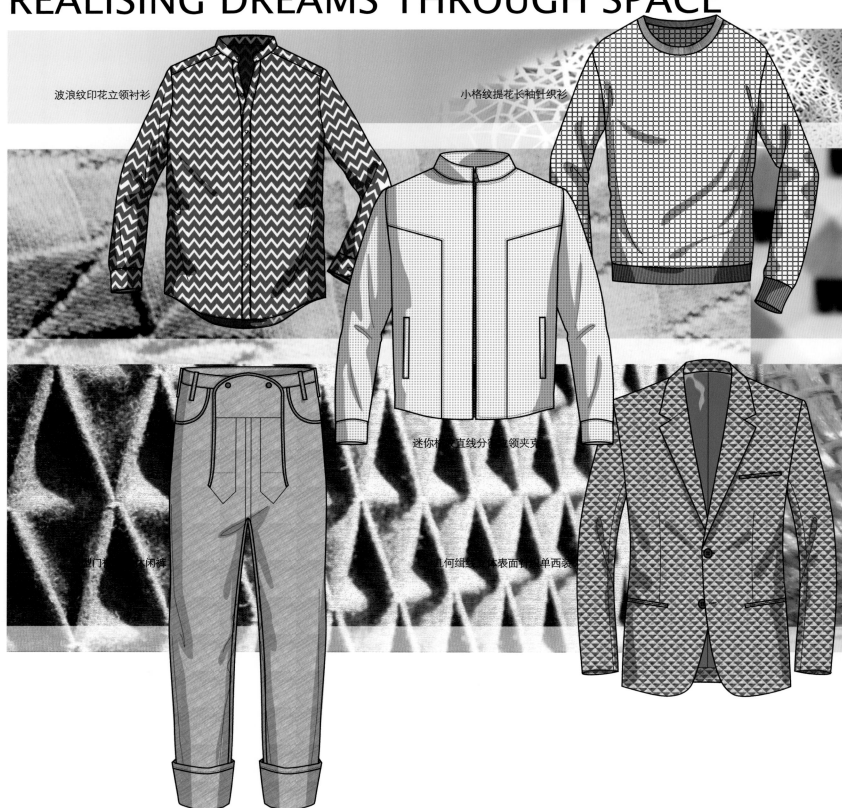

波浪纹印花立领衬衫

小格纹提花长袖针织衫

迷你格纹直线分割立领夹克

门襟休闲裤

几何组织立体表面针织单西装

2015 SPRING/SUMMER STYLE SHANGHAI FASHION TREND / MENSWEAR

关键要素 | 复制粘贴
KEY POINTS | REPETITION & MODULIZATION

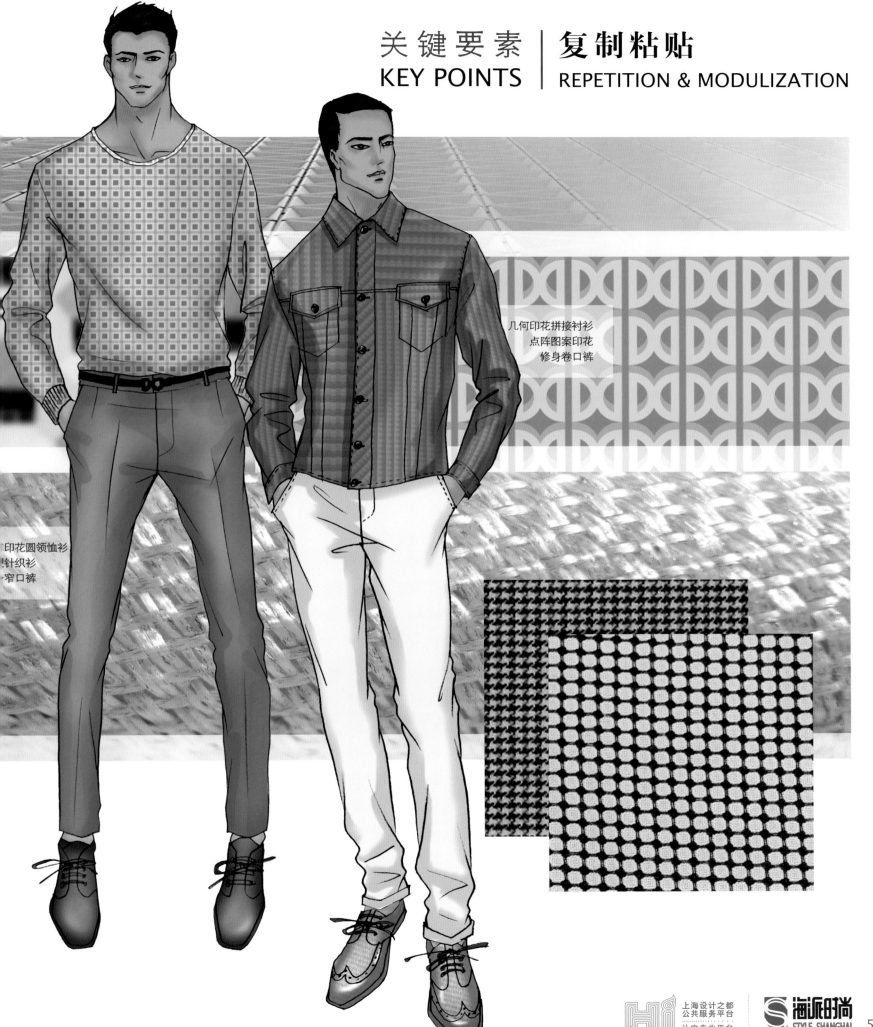

几何印花拼接衬衫
点阵图案印花
修身卷口裤

印花圆领恤衫
针织衫
窄口裤

S男装 2015春夏海派时尚流行趋势

时空筑梦
REALISING DREAMS THROUGH SPACE

数字世界的有序排列

人在网络世界中往往用一个符号表示

网状扫描线剖析人的

白色简洁风衣
满地字母印花圆领恤衫
麻质休闲裤

工作、生活、娱乐都离不开网络这一纽带

设计灵感

网络时代电子产品成为了人与人之间交流的基本方式，在不经意间我们也将成为生活中的一个符号。

In network era, electronic products have become a basic form of communication between people and inadvertently. We will also become a symbol of life.

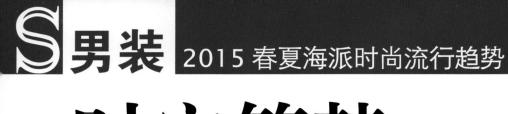

时空筑梦
REALISING DREAMS THROUGH SPACE

双层假领套头条纹针织衫

几何分割圆领长袖针织衫

短款暗门襟造型领单西装

迷你点阵纯色长袖POLO衫

几何印花长袖衬衫

S男装 2015春夏海派时尚流行趋势

时空筑梦
REALISING DREAMS THROUGH SPACE

机械鱼装置设计

海底仿生造型科学实验站

动物纹样印花衬衫
薄型面料
黑色九分裤

从自然界寻找生物的奥秘进行仿生设计

设计灵感

仿生学将科技、自然和人三者融为一体，从生物界中汲取创作灵感，借鉴造物者的杰作，创造出具有生物有机形态的超凡作品。

Bionics integrates science & technology with nature and people. Designers get inspiration from the biosphere, and draw extraordinary works in bio-organic forms with the help from the Creators.

www.style.sh.cn

时空筑梦
REALISING DREAMS THROUGH SPACE

解构仿生分割衬衣　　仿动物表皮光泽感提花针织衫

光泽感平驳领单西

海豚印花短袖T恤　　拼色网眼运动衫

S 男装 2015 春夏海派时尚流行趋势

时空筑梦
REALISING DREAMS THROUGH SPACE

极致统一的家居设计

纯白骨瓷花瓶摆件凸显极简生活态度

大型创意白天鹅装置展览作品

半透明材质套装
夸张的领部、袖部造型
透叠效果

设计灵感

极简设计正顺应潮流展现最单纯、秩序、规则的形态，大块的色彩、纯粹的设计、硬朗的直线条打造未来感极简装束。

Minimalism is becoming the trend by showing the most simple, orderly and regular shapes. Blocks of colors, bold designs and simply straight lines are all elements in futuristic minimalist attire.

S男装

2015 春夏海派时尚流行趋势

时空筑梦
REALISING DREAMS THROUGH SPACE

小翻领暗门襟极简衬衫

双层V领薄型针织衫

一粒扣极窄驳领单西

小立领一字分割夹克

直筒七分休闲裤

2015 SPRING/SUMMER STYLE SHANGHAI FASHION TREND / MENSWEAR

关键要素 | 单纯设计
KEY POINTS | SIMPLE DESIGN

简约极简风衬衣

光泽感材质
极简兜帽夹克衫
修身窄口裤

光色纯棉材质
下装同色搭配
闲短裤

多彩享梦
ENJOY COLORFUL DREAMS

街头亚文化，怪诞风格，童年拾趣，缤纷的色彩，不同元素、概念、风格的组合在玩乐中演绎出新创意，追求感官刺激，享乐生活，成为当代潮人 STYLE。

Combination of elements from street subcultures, grotesque style, childhood memories and various colors has sparkled new ideas in contemporary life style. More and more people are becoming to pursue sensory stimulation and enjoy life in innovative ways.

MENSWEAR
2015 SPRING/SUMMER STYLE SHANGHAI FASHION TREND
海派男装流行趋势

2015 SPRING/SUMMER STYLE SHANGHAI FASHION TREND / MENSWEAR

关键要素 / KEY POINTS

童年拾趣
CHILDHOOD REFLECTIONS

跨界实验
TRANSBOUNDARY EXPERIMENTS

快乐崇拜
JOLLY IDOLATRY

都市街头
CITY STREETS

蜜桃粉
浅裸粉
幻彩红
石榴红
霓虹绿
迎春黄
金币黄
玛瑙蓝
松石绿
冰山蓝

S男装 2015春夏海派时尚流行趋势

多彩享梦
ENJOY COLORFUL DREAMS

童年的玩具和游戏成为怀旧设计的灵感来源

和小伙伴们一起玩乐的时光是童年最美好的回忆

多彩有趣的玩具伴随我们一起度过天真浪漫的童年生活

任天堂等经典游戏图案印花T恤
纯色休闲短裤

设计灵感

游戏、玩具、卡通片伴随我们度过了天真无邪的孩提时代。超级玛丽、变形金刚、童趣涂鸦摇身一变成为了服装上的多彩图案,让我们重拾儿时的趣味印象。

Games, toys and cartoons accompanied us through happy childhood. Super Mario, Transformers and playful graffiti can all be moved onto clothing as colorful patterns, letting us recall sweet memories of childhood.

多彩享梦
ENJOY COLORFUL DREAMS

2015 SPRING/SUMMER STYLE SHANGHAI FASHION TREND / MENSWEAR

关键要素 | 童年拾趣
KEY POINTS | CHILDHOOD REFLECTIONS

随意的大块面色彩拼接
色块拼接提花针织衫
麻质窄口裤

涂鸦风格印花T恤

多彩的趣味图案纹样
经典卡通片角色形象图案衬衫
纯色西短裤

 男装 2015 春夏海派时尚流行趋势

多彩享梦
ENJOY COLORFUL DREAMS

抽象派风格在现代建筑中的体现

奢侈品品牌路易·威登与街头艺术家跨界合作的丝巾设计

反穿效果
拼接拉毛丹宁
宽松休闲短裤

突破传统维度概念的怪诞风格建筑

设计灵感

不同元素、概念、风格的组合在玩乐中演绎出新创意，这些风格迥异的元素搭配打破传统界限，打造出夸张、怪诞的设计风格。

Mixing different elements, concepts and styles together can have chemical reactions and new sparks of ideas, breaking traditional boundaries and creating exaggerated, grotesque styles.

多彩享梦
ENJOY COLORFUL DREAMS

2015 SPRING/SUMMER STYLE SHANGHAI FASHION TREND / MENSWEAR

关 键 要 素 | 跨 界 实 验
KEY POINTS | TRANSBOUNDARY EXPERIMENTS

蕾丝、丝绒等女性化材质应用
女士内衣风格的男士马甲
极简休闲西服，卷口休闲裤

光泽防水雨衣材质的休闲单西
荧光色衬衫
直筒休闲裤

S 男装 2015春夏海派时尚流行趋势

多彩享梦
ENJOY COLORFUL DREAMS

迷幻霓虹灯光营造的欢乐气氛

幽默有趣的元素与缤纷色彩打造欢乐生活

大块面、高饱和度的色彩应用
高饱和度纯色衬衫
亮色领带

沉迷于欢乐多彩的氛围中忘记一切烦恼尽情享乐

设计灵感

仿佛所有多彩、幽默的事物都能吸引潮人们的驻足,带有自嘲意味来体验这些新奇事物,追求感官刺激,享乐生活,成为当代潮人领袖。

It seems that all colorful or humorous things can attract trendsetters' eyes. They'd like to experience the novelty and enjoy life via exaggerative and ironical expression and sensory stimuli.

多彩享梦
ENJOY COLORFUL DREAMS

幽默图案印花恤衫

多彩几何纹样长袖针织衫

万花筒效果迷幻印花夹克

民族印花风格休闲裤

彩色条纹夹克

2015 SPRING/SUMMER STYLE SHANGHAI FASHION TREND / MENSWEAR

关键要素 | 快乐崇拜
KEY POINTS | JOLLY IDOLATRY

大块面不规则撞色几何纹样
幽默图案及文字的应用
宽松休闲裤

多彩民族印花纹样
万花筒般的幻彩图形针织衫
麻质格纹休闲裤

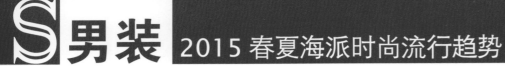

S 男装 2015 春夏海派时尚流行趋势

多彩享梦
ENJOY COLORFUL DREAMS

上海 M50 创意园区外墙涂鸦艺术作品

源自美国黑人文化的 Hip-Hop，通过舞蹈、服饰等对当代青年人有着深刻影响

多彩霓虹灯营造的

大面积趣味性波普图案
波普图案印花 T 恤
残破做旧牛仔裤

街头涂鸦、极限运动、摇滚乐是青年人宣泄自我情绪的渠道

设计灵感

涂鸦、霓虹、朋克、波普、摇滚音乐等展现都市街头亚文化，展现当代青少年追求自我实现的生活态度。

Graffiti, neon, punk, pop & rock music are all parts of street sub-culture, indicating young people's pursuit of self-realization in urban life.

2015 SPRING/SUMMER STYLE SHANGHAI FASHION TREND / MENSWEAR

关键要素 KEY POINTS | 都市街头 CITY STREETS

- 浅裸粉
- 珊瑚红
- 深桃红
- 泳池蓝
- 玛瑙蓝
- 松石绿
- 湖泊蓝

街头风格贴布绣

趣味迷彩图案

骷髅、手枪等街头元素图案装饰
大面积拼色针织衫
亮色针织衫

图案兜帽风衣
衬衫
卷口牛仔裤

外墙上的涂鸦作品

多彩享梦
ENJOY COLORFUL DREAMS

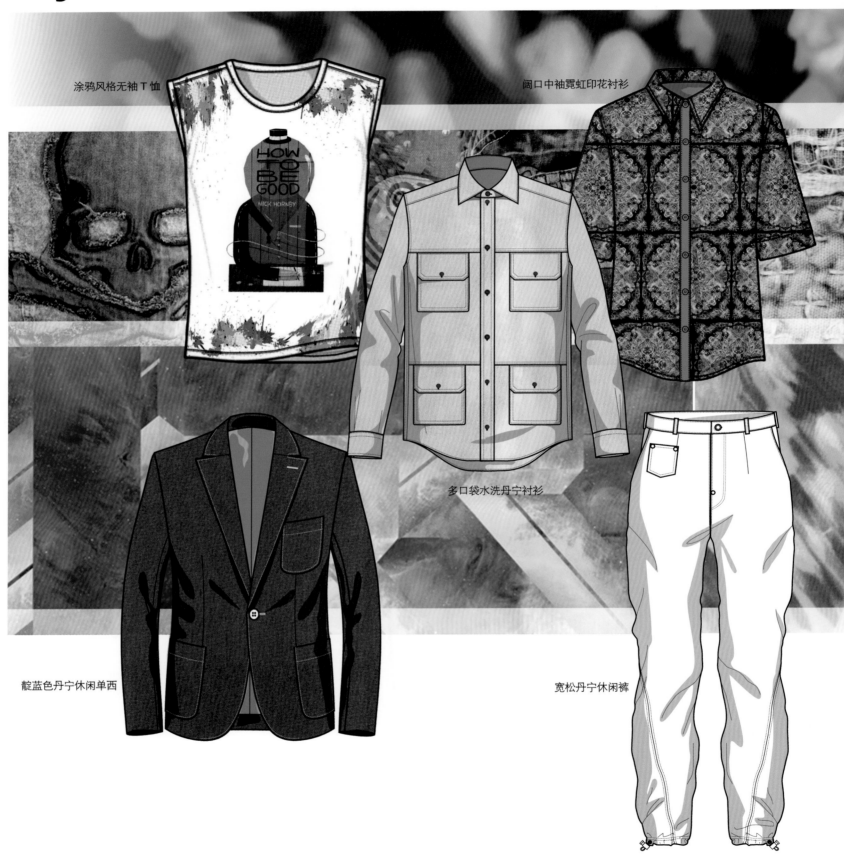

2015 SPRING/SUMMER STYLE SHANGHAI FASHION TREND / MENSWEAR

关键要素 | 都市街头
KEY POINTS | CITY STREETS

光泽感面料
蕾丝等女性化面料
光泽印花夹克
光泽感直筒裤

色系迷彩图案夹克
鸟风格印花T恤
身休闲裤

女装
WOMENSWEAR

| 海上星梦
STAR DREAMS OVER THE SEA
优活之梦
LIVING THE DREAM
时空筑梦
REALISING DREAMS THROUGH SPACE
多彩享梦
ENJOY COLORFUL DREAMS

海上星梦
STAR DREAMS OVER THE SEA

夜晚，梦搅乱心头，如星光的思绪在弥漫。推开临江的花窗，守望时光的涟漪，用一世沉淀撰写岁月的记忆。曾经的繁华风姿被重新演绎，东方的文化与西方的古典细节完美统一，绽放出美丽与精彩。

When night falls and dreams let us explore different virtual worlds. Open traditional lattice windows and think back to memorable sweetness and bitterness. Old stories have been rewritten and old bustling charms of this city have been refurnished. When oriental culture and Western classical details fuse with each other, a new beautiful and wonderful combination will present itself.

WOMENSWEAR

2015 SPRING/SUMMER STYLE SHANGHAI FASHION TREND
海派女装流行趋势

2015 SPRING/SUMMER STYLE SHANGHAI FASHION TREND / WOMENSWEAR

关键要素 | KEY POINTS

繁华时代
FLOURISHING ERA

精细品质
EXQUISITE QUALITY

新贵气质
NOBLE TEMPERAMENT

创新传统
INNOVATING TRADITIONS

色号
浪花白
青釉灰
青砖灰
杏仁黄
琥珀橙
皮革黄
胭脂红
酸枝红
紫檀棕
乌梅黑

S 女装 2015春夏海派时尚流行趋势

海上星梦
STAR DREAMS OVER THE SEA

华丽刺绣的衣领

20世纪30年代电影场景剪影　　　　　　　　　　　　　木质镂空窗户细节

百乐门的歌舞升平　　　　　　　珍珠、水晶耳环配饰

设计灵感

如花美眷，似水流年。耳畔流过轻声细语，心上飘过绰约影姿。一抹芳香的回忆久久停留在回眸人的心头。繁华落尽于海上，再一次绽放迷人魅力。

Beauties are like flowers as youth passing as a fleeting wave. Flowing through ears are those mellifluous whispers, staying in hearts are those unforgettable graceful postures and traces of fragrance. This prosperous city opens her arms once again with her blooming charm.

2015 SPRING/SUMMER STYLE SHANGHAI FASHION TREND / WOMENSWEAR

关键要素 | 繁华时代
KEY POINTS | FLOURISHING ERA

的手工刺绣

藤蔓状的珠绣细节
半透明的蕾丝与塔夫绸面料
镂空透叠的效果

细密繁复的蕾丝
尤雅的线条
飘逸的拖地裙摆
不旧的复古

舞台模糊的光线

刺绣手包

浪花白
青釉灰
杏仁黄
琥珀橙
胭脂红
酸枝红
瓷瓦黑

海上星梦
STAR DREAMS OVER THE SEA

纤细轻巧的装饰细节

复古雕花装饰

细腻的纹理

贝壳形态饰件

设计灵感

精灵般轻巧的步伐，似珍珠般散发着柔软细腻的光晕。不错过生活中每一个精致的细节，在纷繁的时光中，从容优雅地前行。

Fair lady walks with light paces and her soft and delicate skin glows like pearls. She'll take good care of every detail and keep an elegant attitude towards this complicated world.

海上星梦
STAR DREAMS OVER THE SEA

S女装 2015 春夏海派时尚流行趋势

海上星梦
STAR DREAMS OVER THE SEA

旋转楼梯细节

精细复古的打字机

卷草纹墙面装饰

欧式浮雕建筑

欧式风格提花餐具

设计灵感

内敛的格调与张扬的优雅，东方文化意蕴与西方古典风情兼具的细节，两种对立感觉完美统一，超越时尚与潮流，正是城中之人追求的高雅生活方式。

Introverted grace and symbolic elegance, cultural implication of the east and classic style of the west, two seeming opposites can be perfectly united. This kind of mixture beyond only fashion or trends is what people living in this city are looking for.

海上星梦
STAR DREAMS OVER THE SEA

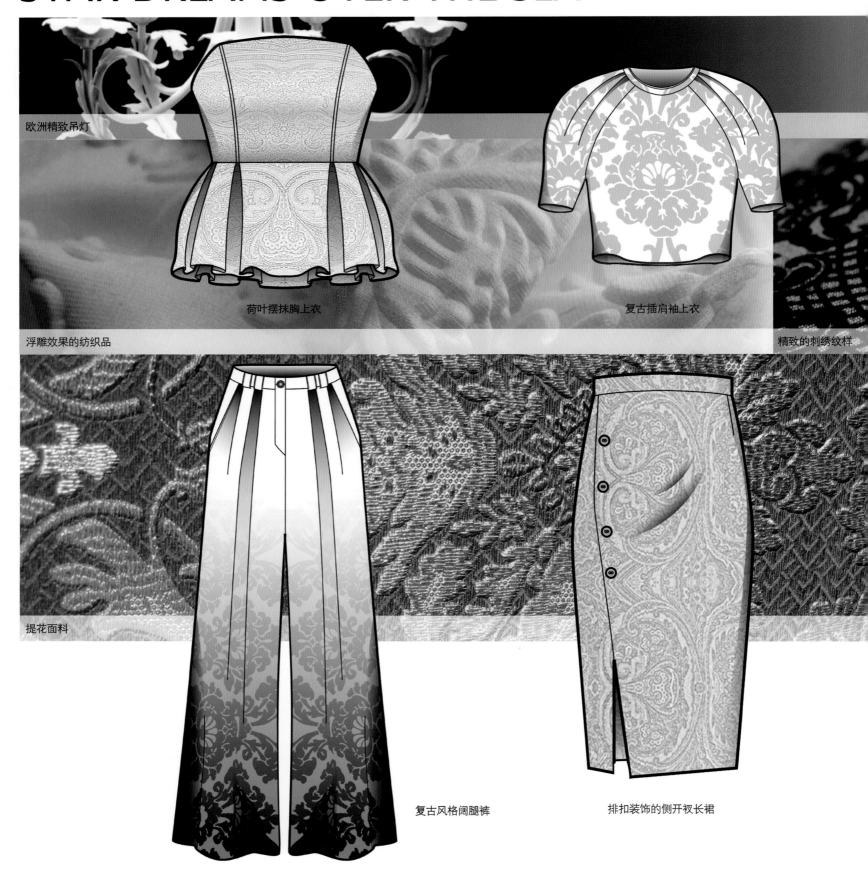

欧洲精致吊灯

荷叶摆抹胸上衣

复古插肩袖上衣

浮雕效果的纺织品

精致的刺绣纹样

提花面料

复古风格阔腿裤

排扣装饰的侧开衩长裙

S 女装 2015春夏海派时尚流行趋势

海上星梦
STAR DREAMS OVER THE SEA

传统元素的现代演绎

传统与现代的时尚碰撞

皮革的创新编织方法

上海电影博物馆装饰细节

传统装饰风格壁柜

设计灵感

新与旧的对撞，历史与时尚的交融，迸发出新的魅力。这座城市曾经的繁华风姿也将被重新演绎。

As the new and the old, history and fashion colliding with each other, fresh charm will be created. A new show of old time prosperity of this city will be on with new elements being added.

2015 SPRING/SUMMER STYLE SHANGHAI FASHION TREND / WOMENSWEAR

关键要素 | 创新传统
KEY POINTS | INNOVATING TRADITIONS

"双妹"丝巾图案

乌梅黑
瓷瓦黑
紫檀棕
酸枝红
皮革黄
杏仁黄
青釉灰

经典的旗袍造型
与利落的分割装饰的完美结合
既有东方的传统韵味
又不失现代女性的风姿

改良了传统的旗袍廓型
加入现代风格的几何形式装饰
大胆的异色拼接
典雅亮丽气质

复古风头饰

"上海滩"手包

海上星梦
STAR DREAMS OVER THE SEA

女装 2015 春夏海派时尚流行趋势

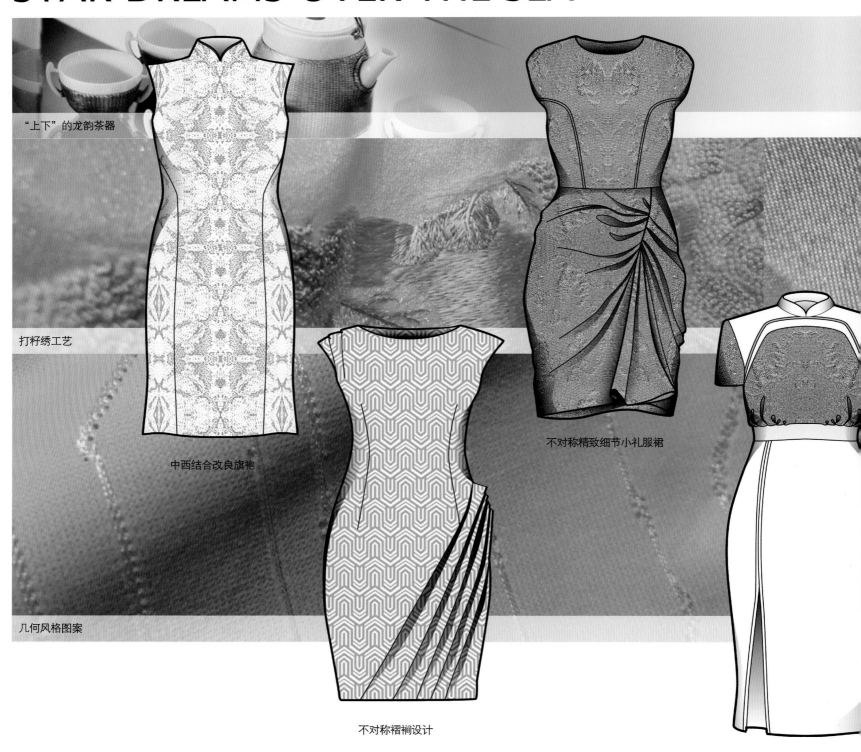

"上下"的龙韵茶器

打籽绣工艺

几何风格图案

中西结合改良旗袍

不对称褶裥设计

不对称精致细节小礼服裙

拼色设计现代旗袍

2015 SPRING/SUMMER STYLE SHANGHAI FASHION TREND / WOMENSWEAR

关键要素 | 创新传统
KEY POINTS | INNOVATING TRADITIONS

中式服装细节

改良唐装设计
加以华丽的刺绣
和现代感强烈的剪裁
搭配西装短裤
优雅而性感

经典斜纹面料

上衣的精致分割线设计
不对称的高开衩半裙
既体现了传统的审美风格
又不失利落与前卫

复古手绘图案

乌梅黑
瓷瓦黑
紫檀棕
酸枝红
皮革黄
杏仁黄
青釉灰

S 女装　2015 春夏海派时尚流行趋势

优活之梦
LIVING THE DREAM

和谐的微风，吹开温暖的心，杨柳吐绿，油菜花香，河清湖秀，这里一片生机盎然。平凡的生活，绿色的梦，拥抱自然，拥抱蓝天，大地景色更新，谱写更美的乐章。

When warm breeze blows into heart, light green buds sprout from willow trees, canola flowers smell so fresh, calm lake surface shimmers under sunshine, you know that spring has come. People living in cities start to construct dreams in green wishing to embrace nature and blue sky and enjoy magnificent scenes.

WOMENSWEAR
2015 SPRING/SUMMER STYLE SHANGHAI FASHION TREND
海派女装流行趋势

2015 SPRING/SUMMER STYLE SHANGHAI FASHION TREND / WOMENSWEAR

关键要素 | KEY POINTS

织补风尚
BACK TO THE BASICS

平民贵族
ORDINARY NOBILITY

生态材质
ECO MATERIAL

家庭手工
FAMILY HANDICRAFTS

淡奶黄
蜜瓜黄
香芋灰
抹茶绿
艾草绿
苔藓绿
水晶蓝
孔雀蓝
青花蓝
枣泥红

S 女装 2015春夏海派时尚流行趋势

优活之梦
LIVING THE DREAM

旧衣新貌，拼接创意

纯手工艺，精工细作

"织布鸟"编织巢穴

多种色彩肌理的拼接

设计灵感

在过去岁月里渐渐淡出的织补，而今又重新走上现代时尚的舞台，成为一种生活态度，一种精神。不拘泥于过去与现在，赋予旧物以新的生命力。

Darn fading gradually in the past years now again embarks on modern fashion stage, becoming an upright attitude to life and a spirit. Not rigidly adhering to the boundary between the past and the present, people are now crossing the line and giving old things with new vitality.

www.style.sh.cn

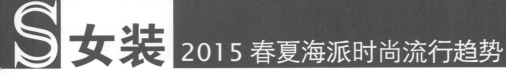

优活之梦
LIVING THE DREAM

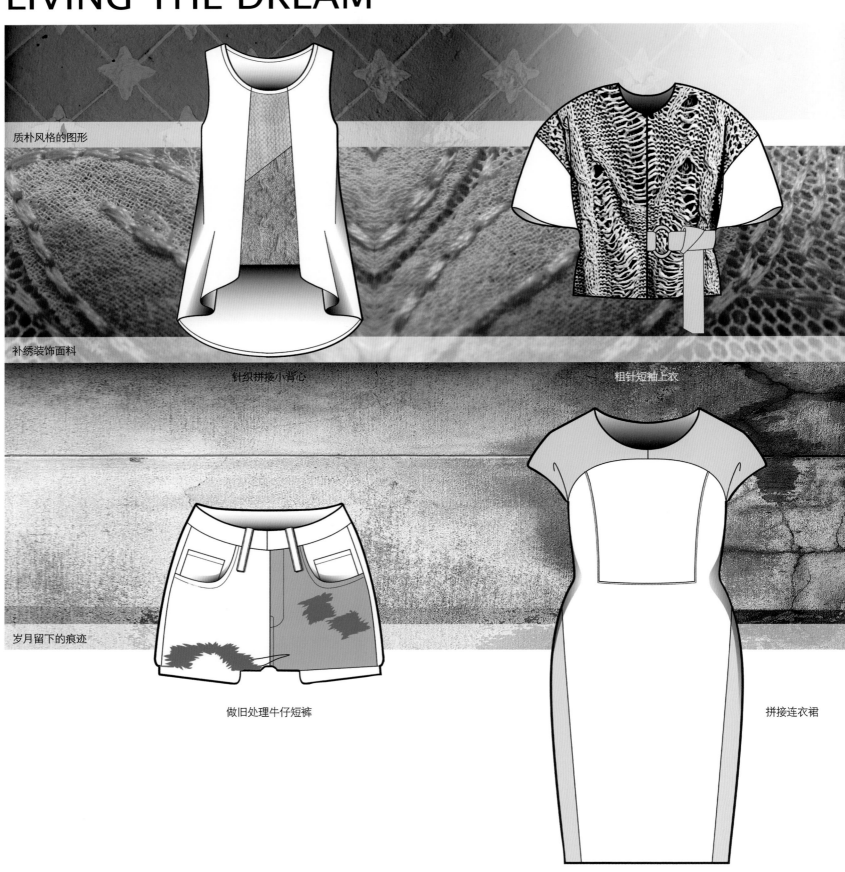

质朴风格的图形

补绣装饰面料

针织拼接小背心

粗针短袖上衣

岁月留下的痕迹

做旧处理牛仔短裤

拼接连衣裙

S 女装 2015春夏海派时尚流行趋势

优活之梦
LIVING THE DREAM

五线谱上的珍珠音符

上海弄堂角落拾景

大线条小细节

经典老式自行车

精致的异色小方领

设计灵感

弄堂里平凡悠闲的生活不乏对高雅和品质的追求。简洁明朗的轮廓，朴实的棉麻质地，精致的细节加上小巧的印花，老上海的韵味弥漫开来。

Alley-life is ordinary and slow-pace but with no sacrifice of elegance and quality. Simple and clear shape, cotton texture, exquisite details coupled with small prints, all these will remind you of the charm of old Shanghai style.

2015 SPRING/SUMMER STYLE SHANGHAI FASHION TREND / WOMENSWEAR

关键要素 | 平民贵族
KEY POINTS | ORDINARY NOBILITY

花白衬衫

…面料的西装外套
…装饰的半袖
…门襟设计
…胸前塔克连身裙

宽松款和服袖上衣
精致蕾丝面料
领子袖口的拼色设计
与西短裤相呼应

蛋壳黄
淡奶黄
蜜瓜黄
香芋灰
水晶蓝
斗笠黄
枣泥红

午后的红茶

雅致图案真丝方巾

优活之梦
LIVING THE DREAM

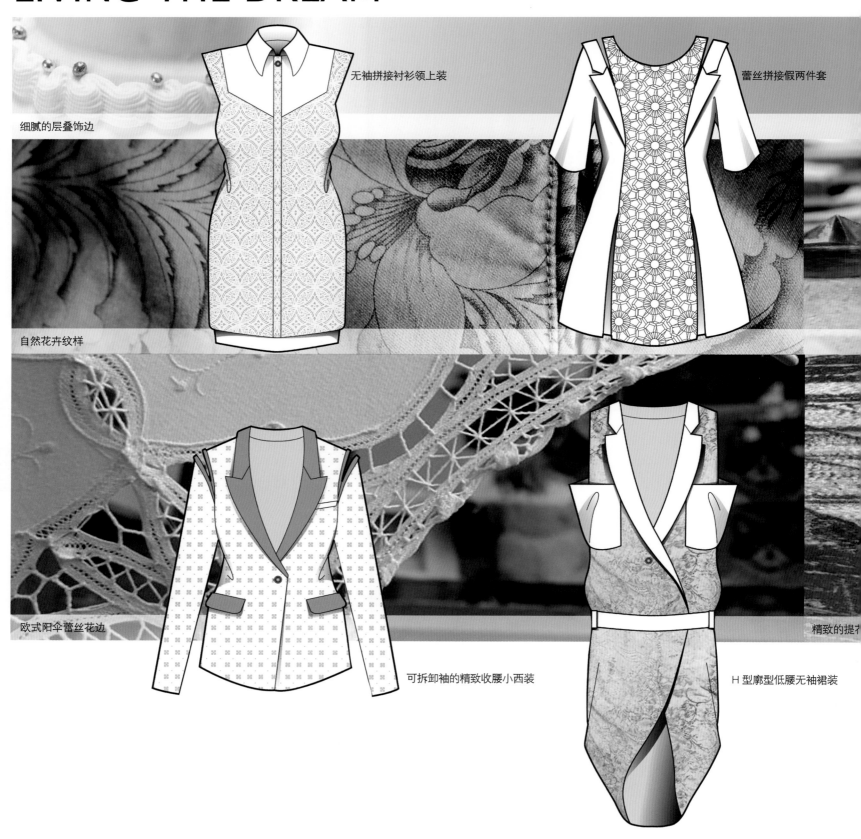

- 细腻的层叠饰边
- 自然花卉纹样
- 欧式阳伞蕾丝花边
- 无袖拼接衬衫领上装
- 蕾丝拼接假两件套
- 可拆卸袖的精致收腰小西装
- H型廓型低腰无袖裙装
- 精致的提花

S 女装 2015 春夏海派时尚流行趋势

优活之梦
LIVING THE DREAM

自然生长的苔藓

棕红色的落叶

盛开的绣球花

晨露中的麦穗

设计灵感

弥漫着泥土芳香的麦田，光洁而凝静的鹅卵石，沉寂的枯叶与轻盈的花瓣合奏出一曲生态和谐的韵律，大自然一派生机盎然。

Wheat fields filled with aroma of soil, clean cobblestones lying still on ground, dead leaves and delicate petals are playing a rhythm in harmony. Getting yourself close to nature and you'll also feel that vitality.

镶嵌天然石的手包

2015 SPRING/SUMMER STYLE SHANGHAI FASHION TREND / WOMENSWEAR

关键要素 | 生态材质
KEY POINTS | ECO MATERIAL

上的露水

短袖长款衬衫
不对称棉麻衣片设计
斜开衩飘逸下摆
随性自然轻松自由

立领小背心
外搭露肩棉布拼色小外套
米色简洁纯棉裤装
充满自然的色调

圆润的鹅卵石

骨骼项链

- 淡奶黄
- 蛋壳黄
- 香芋灰
- 水晶蓝
- 红雀蓝
- 艾草绿
- 苔藓绿

优活之梦
LIVING THE DREAM

干裂的大地

枯叶纹理

岩石表面肌理图案

自然花卉图案双排扣外套

仿鳄鱼皮短外套

印花背心式抽褶短裙

斑驳肌理抹胸裙

|S| 女装 2015 春夏海派时尚流行趋势

优活之梦
LIVING THE DREAM

手工珠绣细节

手工编织杯垫和银汤匙　　　　　　　　　　　　　　家庭手工

编织小竹篮

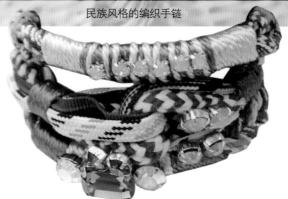
民族风格的编织手链

设计灵感

拒绝信息化社会产品的冰冷感觉，期待凝固着温暖人情味的手工作品，细致情感与粗糙质地带来了独特的怀旧风韵。

Say no to the cold feelings coming as part of the information society and look forward to handiworks embodying humanity feelings. Combination of meticulous emotion and rough texture brings unique nostalgic charm.

2015 SPRING/SUMMER STYLE SHANGHAI FASHION TREND / WOMENSWEAR

关键要素 | 家庭手工
KEY POINTS | FAMILY HANDICRAFTS

景泰蓝装饰品

针织小套装
强调粗细纹理拼色效果
裙摆的针织与梭织拼接
衣身前片夸张的绞花设计
独特而迷人

款棉麻休闲小外套
落肩和悬垂门襟设计
滚边装饰
几何图案合身恤衫
瘦腿牛仔裤

染色毛线　　　　　　编织表带的手表

艾草绿
水晶蓝
红雀蓝
青花蓝
枣泥红
斗笠黄
紫砂棕

优活之梦
LIVING THE DREAM

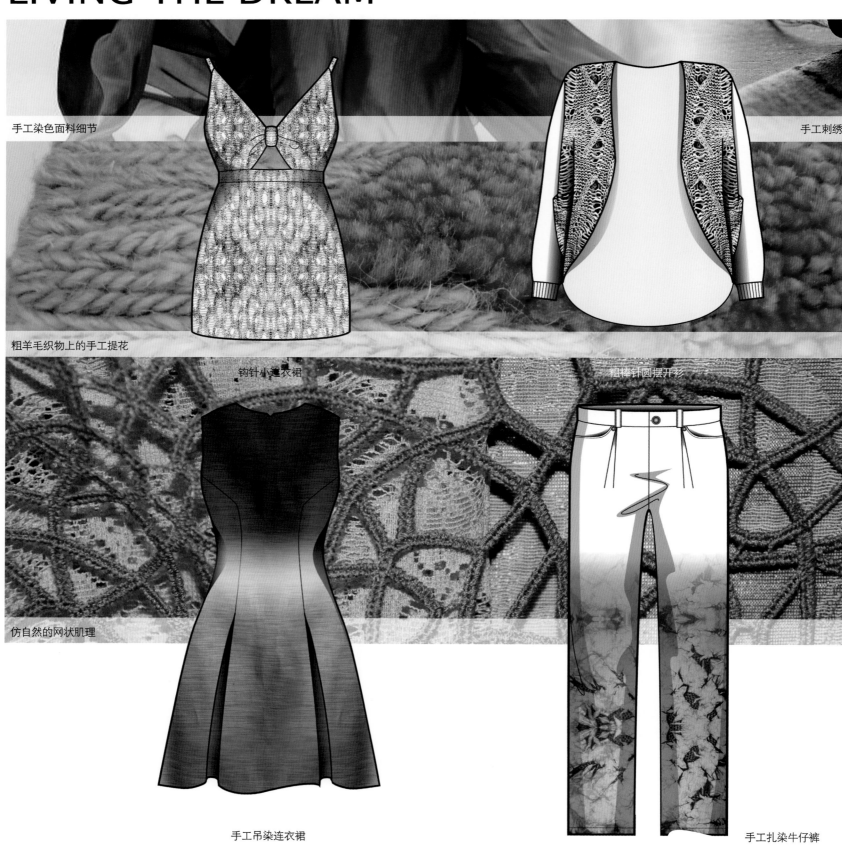

手工染色面料细节

手工刺绣

粗羊毛织物上的手工提花

钩针小连衣裙

粗棒针圆摆开衫

仿自然的网状肌理

手工吊染连衣裙

手工扎染牛仔裤

S 女装 2015 春夏海派时尚流行趋势

时空筑梦
REALISING DREAMS THROUGH SPACE

高科技时代,生物机械化显现,现代人思维方式被电脑同化,被符号固化,不断以复制粘贴的形式重复……时光如梭,不变的是纯净的内心所构筑的梦想,未来归于最单纯的自我。

Because of huge impact brought by high-tech and the appearance of biological mechanization, the ways of how different people think tend to assimilate with each other via frequently quoting symbols, using computers and "copy and paste". Time flies, the only thing that hasn't changed is the dream from the bottom of heart, that we can be the purest selves in future.

WOMENSWEAR
2015 SPRING/SUMMER STYLE SHANGHAI FASHION TREND
海派女装流行趋势

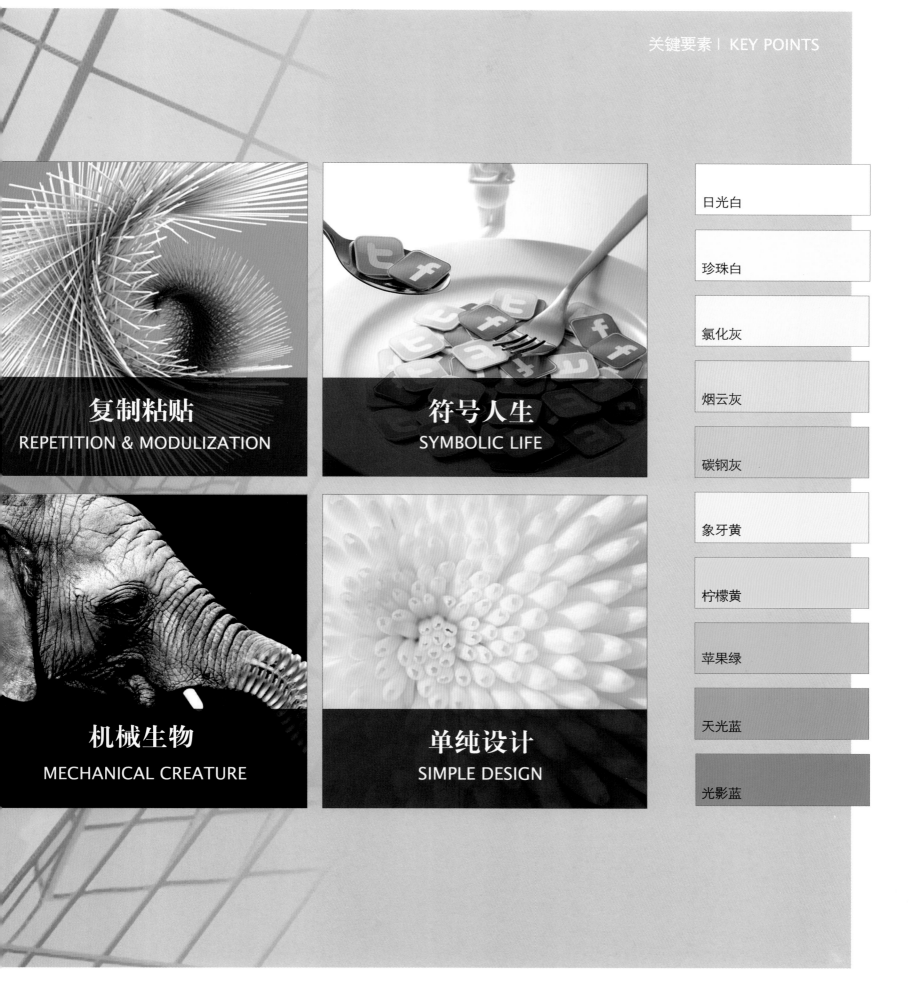

女装 2015 春夏海派时尚流行趋势

时空筑梦
REALISING DREAMS THROUGH SPACE

在复制粘贴中挖掘香氛般的欢愉

梦幻透彻的蒲公英　　　　　　　　　　　　　　　　　重复效果的玻璃墙面

克隆人中关于基因复制的应用

设计灵感

一个个分子组成浩瀚的宇宙，最最微小的细胞组成复杂的器官。在不断的重复前进中寻找发展的根源，举一反三的精神需要传承更需要被打破。

The universe is composed of countless molecules and the most complex organ cannot be in existence without the tiniest cells. We need to find the thruster of development during the process of repetition.

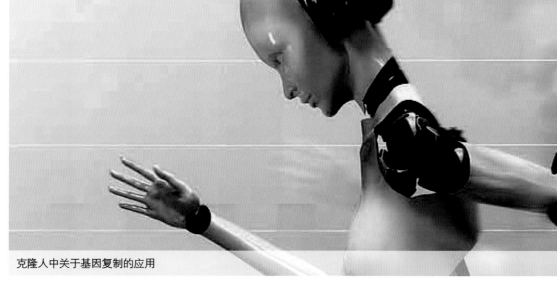

重复棱角拼接艺术沙发

时空筑梦
REALISING DREAMS THROUGH SPACE

DNA 遗传密码中的基因复制

针织镂空面料

层叠门襟上衣

几何重复图案印花衬衫

重复印花的面料

多块面分割拼接短裙

视错觉叠加连体衣

时空筑梦
REALISING DREAMS THROUGH SPACE

古老的中国文化中有许许多多的符号

一个简单的"@"，锁住了多少朋友见面的畅谈。

美味也可以成为符号

未来感的自行车停车标识

具有符号象征意义的饰品

设计灵感

一个图形、一个动作、一个色彩可以使语言简化，沟通更简单，那些本无意义的符号已深深刻入时空的光纤，穿梭在人们的生活意识之中。

A graphic, an action and a color can tell a lot of things instead of speaking them out. Communication has been made much easier than you think. Those once meaningless symbols have been running in optical fibers and making large impact on people's lives.

时空筑梦
REALISING DREAMS THROUGH SPACE

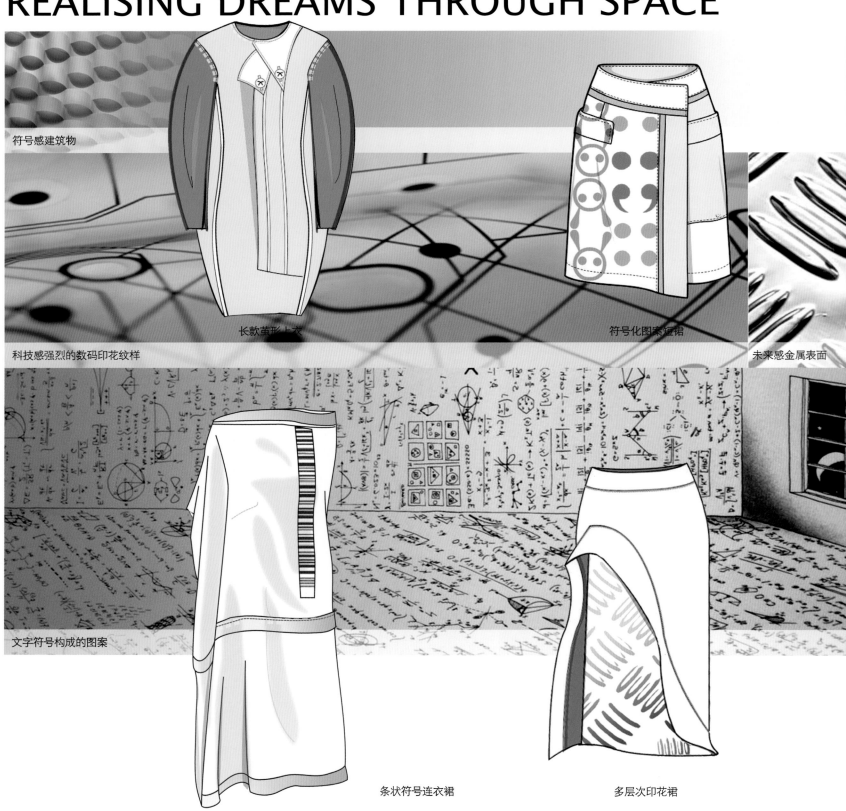

符号感建筑物

长款茧形上衣　　符号化图案短裙

科技感强烈的数码印花纹样　　未来感金属表面

文字符号构成的图案

条状符号连衣裙　　多层次印花裙

女装　2015春夏海派时尚流行趋势

时空筑梦
REALISING DREAMS THROUGH SPACE

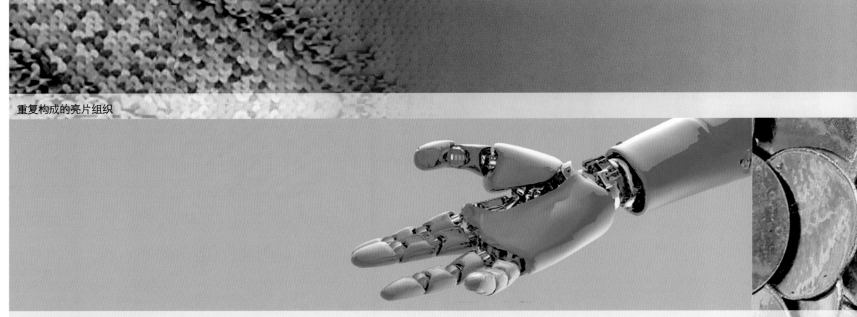

重复构成的亮片组织

未来感高科技机械手臂局部　　　　　　　　　　　　　　　　　　　高纯度金属材质肌理

生物组构的机械形态

超现实直线型

设计灵感

简化的思维，机械的生产。高科技风潮植入生活，机械元件组成新的生物体，它们是四维空间中强烈独特的组织结构。

They have simple minds and can be made in modern mass production. Life embraces technologies in many aspects. Mechanical components compose new organisms. They are unique structures in the 4D world.

时空筑梦
REALISING DREAMS THROUGH SPACE

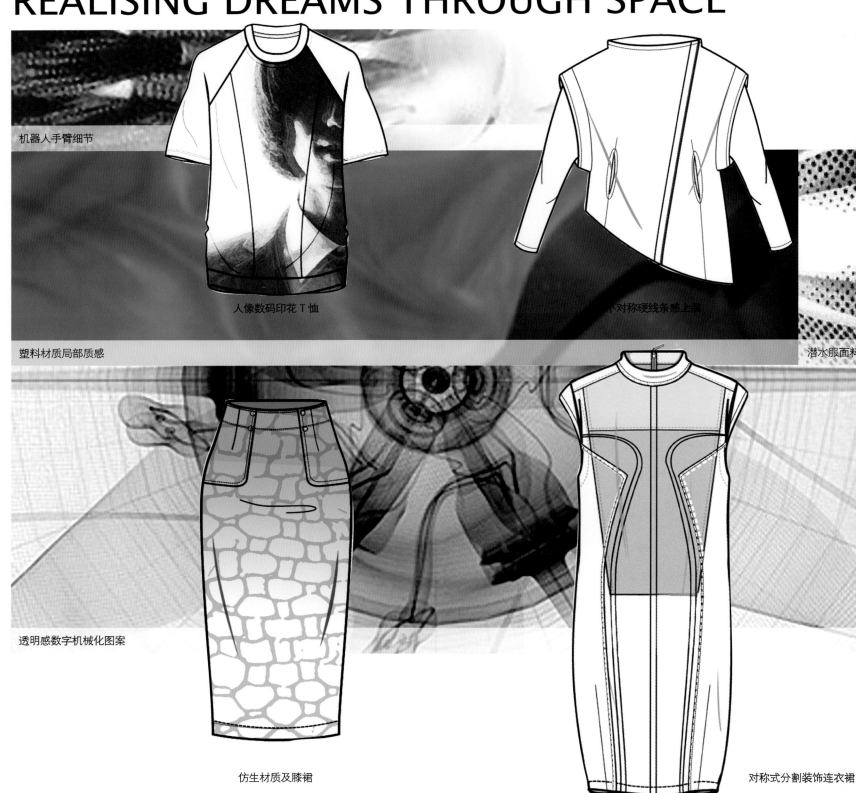

时空筑梦
REALISING DREAMS THROUGH SPACE

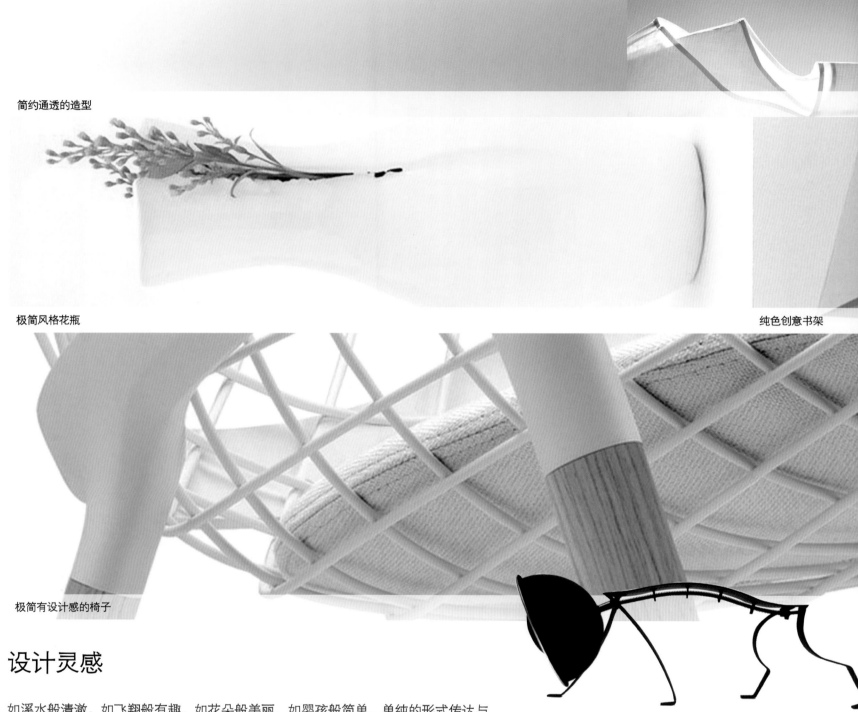

简约通透的造型

极简风格花瓶

纯色创意书架

极简有设计感的椅子

线条感台灯

设计灵感

如溪水般清澈，如飞翔般有趣，如花朵般美丽，如婴孩般简单，单纯的形式传达与讲述，净化着人们生活的烦杂，让浮夸归于零。

As clear as streams, as interesting as flying, as beautiful as flowers, as pure as babies, simple forms help us convey to people our desire for minimalism and unsophisticated life.

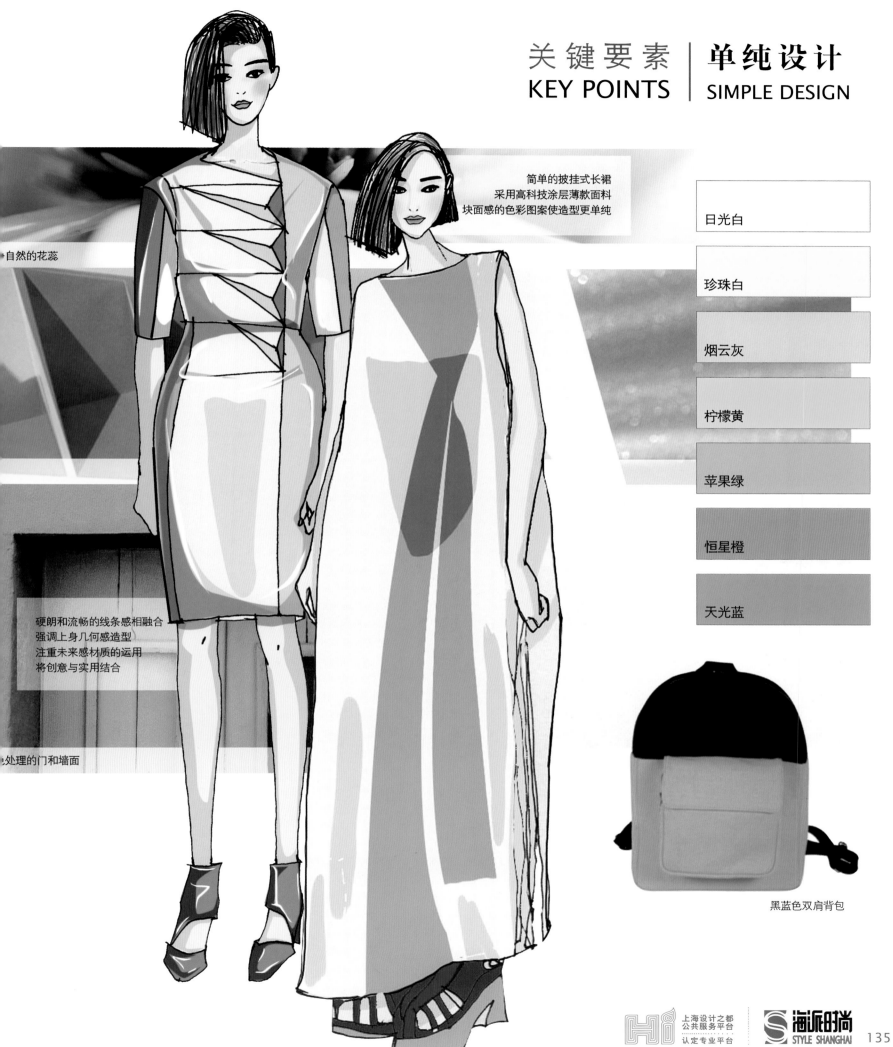

时空筑梦
REALISING DREAMS THROUGH SPACE

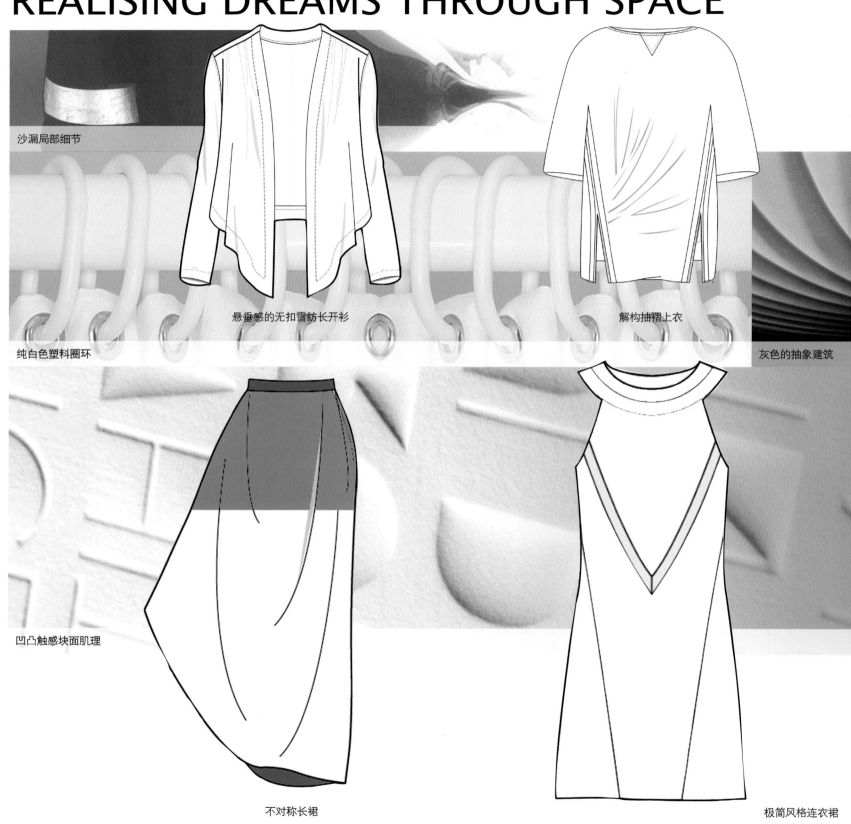

沙漏局部细节

悬垂感的无扣雪纺长开衫

解构抽褶上衣

纯白色塑料圈环

灰色的抽象建筑

凹凸触感块面肌理

不对称长裙

极简风格连衣裙

女装 2015 春夏海派时尚流行趋势

多彩享梦
ENJOY COLORFUL DREAMS

高楼林立，车水马龙的大都市，熙来攘往人群如潮。缤纷的色彩，跳跃的音符，童年的趣事，梦中的欢笑，跨界的交融，亦幻亦真，恍如身处奇妙的梦幻时空。

This city is so colorful and noisy, composed of magnificent skyscrapers and streets crowded with people and vehicles. Designs inspired by childhood and dreams reminding you good old days sometimes make you feel confused about reality and fantasy.

WOMENSWEAR
2015 SPRING/SUMMER STYLE SHANGHAI FASHION TREND
海派女装流行趋势

2015 SPRING/SUMMER STYLE SHANGHAI FASHION TREND / WOMENSWEAR

关键要素 | KEY POINTS

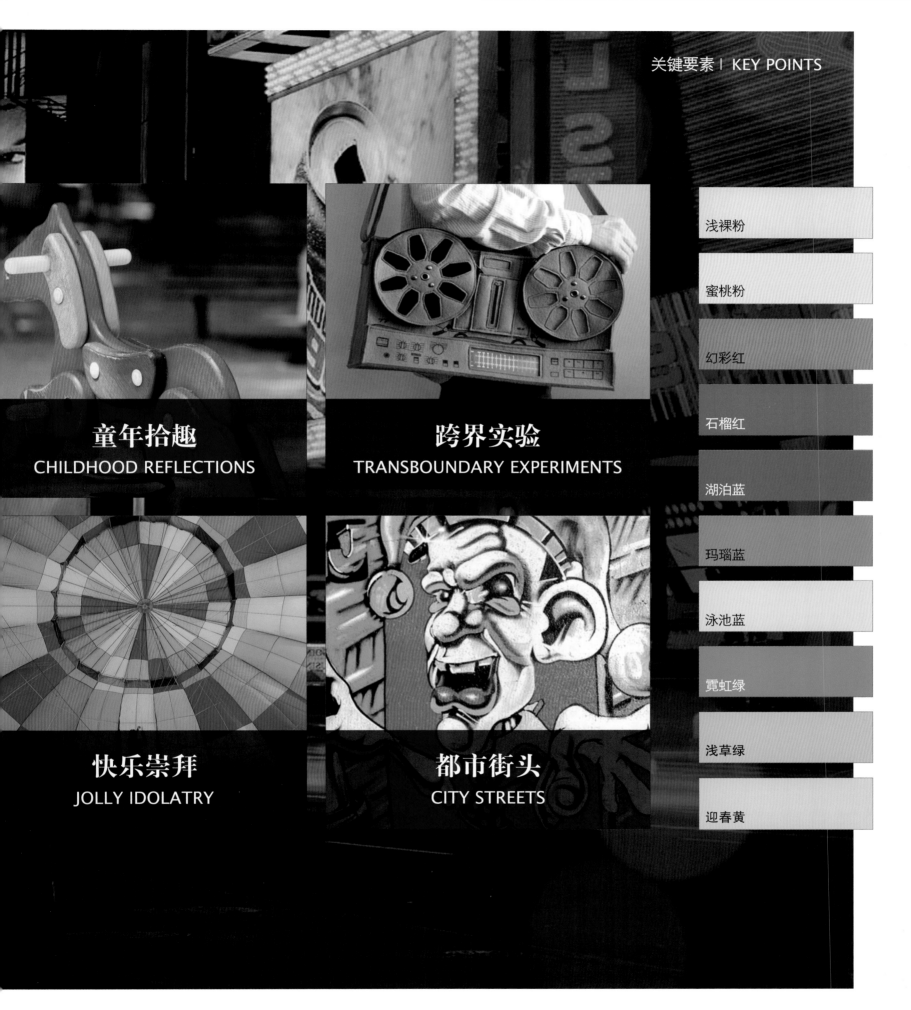

童年拾趣
CHILDHOOD REFLECTIONS

跨界实验
TRANSBOUNDARY EXPERIMENTS

快乐崇拜
JOLLY IDOLATRY

都市街头
CITY STREETS

浅裸粉
蜜桃粉
幻彩红
石榴红
湖泊蓝
玛瑙蓝
泳池蓝
霓虹绿
浅草绿
迎春黄

S 女装 2015 春夏海派时尚流行趋势

多彩享梦
ENJOY COLORFUL DREAMS

手工DIY配饰

记忆中的童年玩具　　　　　　　　　　　　　　　　　　　　　　色彩丰富的光感材

玩具的抽象形态集合

老鼠针包

设计灵感

梦中的旋转木马充满欢笑，甜蜜的棒棒糖捧在手心，灰姑娘最后嫁给白马王子，这一切如蒲公英般快乐的散开，梦想如童话般美好！

Childhood dreams were full of laughter: riding on carousel, holding a sweet lollipop in hand. Happiness spreads like dandelion flying in wind. We all believe that our dreams will come true just as the end of the fairy tale – Cinderella finally lives happily together with Prince Charming.

多彩享梦
ENJOY COLORFUL DREAMS

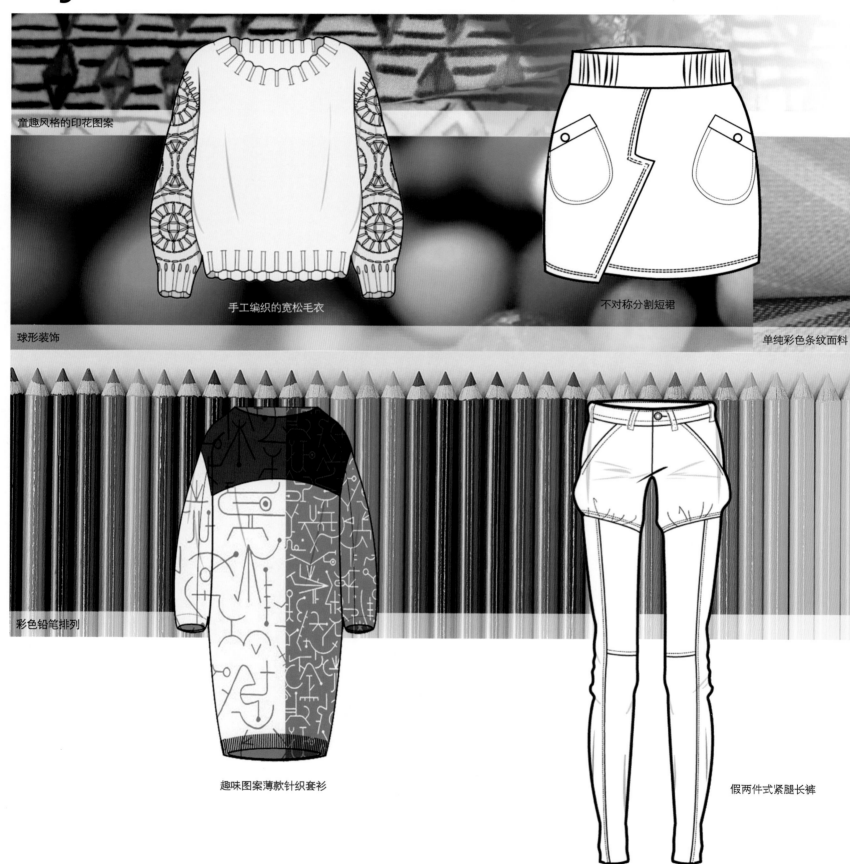

童趣风格的印花图案

手工编织的宽松毛衣

球形装饰

不对称分割短裙

单纯彩色条纹面料

彩色铅笔排列

趣味图案薄款针织套衫

假两件式紧腿长裤

S 女装 2015 春夏海派时尚流行趋势

多彩享梦
ENJOY COLORFUL DREAMS

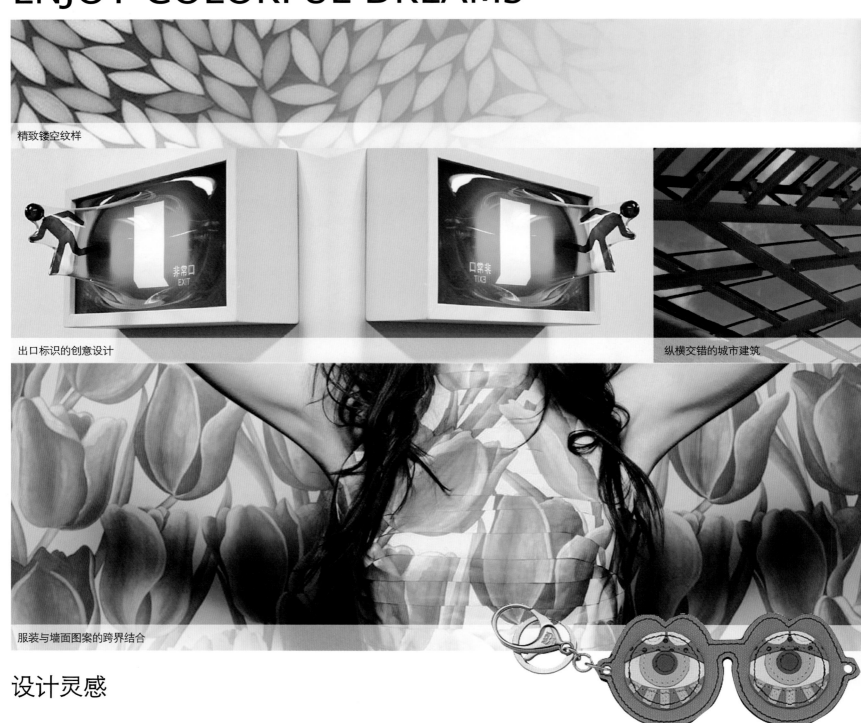

精致镂空纹样

出口标识的创意设计

纵横交错的城市建筑

服装与墙面图案的跨界结合

"MCM"品牌与当代流行艺术家合作玩味眼镜

设计灵感

原本毫不相干，甚至矛盾、对立的元素碰撞出灵感火花，相互渗透，转折打破，一种意料之外的融合，成就一个奇妙的全新梦想。

Irrelevant, even contradictory, opposing elements will collide with each other creating sparks of inspiration. They penetrate into the other, change themselves and break boundaries. The unexpected fusion will bring us a wonderful brand-new dream.

2015 SPRING/SUMMER STYLE SHANGHAI FASHION TREND / WOMENSWEAR

关键要素 | 跨界实验
KEY POINTS | TRANSBOUNDARY EXPERIMENTS

H型宽松简洁廓型
挺括的钟形连袖
底摆吊染刷色渐变
视觉中心集中上移

冰山蓝
泳池蓝
玛瑙蓝
松石绿
蜜桃粉
浅裸粉
琵琶橙

安迪·沃霍尔丝网印玩偶

多彩享梦
ENJOY COLORFUL DREAMS

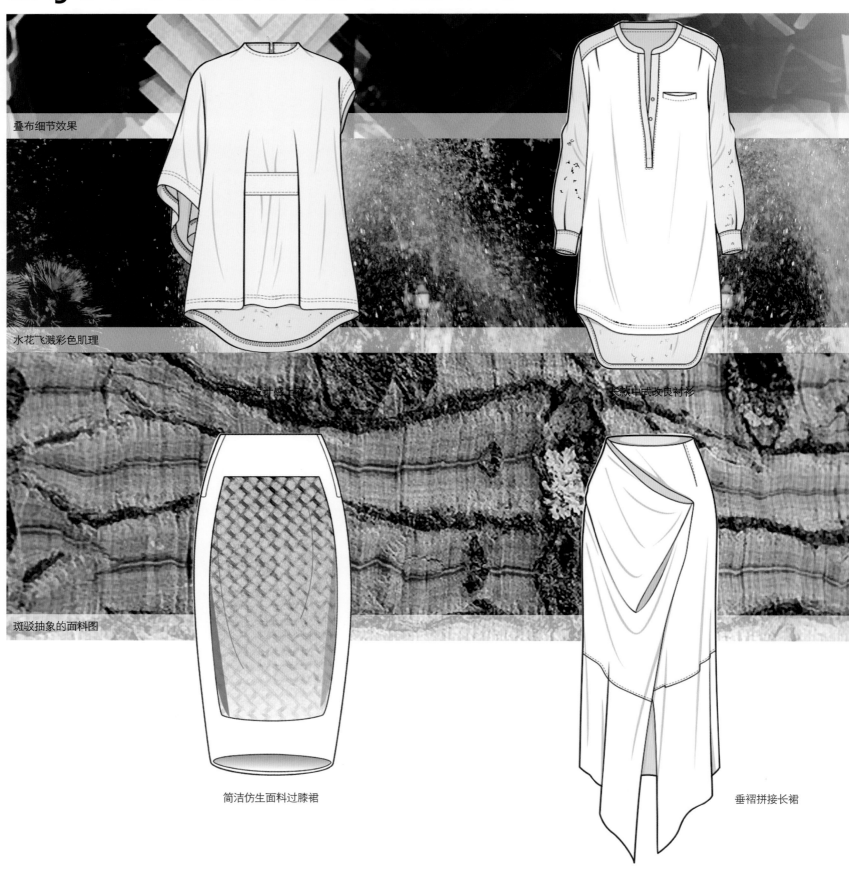

叠布细节效果

水花飞溅彩色肌理

斑驳抽象的面料图

简洁仿生面料过膝裙

垂褶拼接长裙

S 女装 2015 春夏海派时尚流行趋势

多彩享梦
ENJOY COLORFUL DREAMS

多彩的纱线

层叠交错的霓虹灯管

展现年轻、欢乐轻松氛围的创意涂鸦墙

多彩时尚的欧式家具

设计灵感

放下所有的包袱，忘记自己的存在，扬起嘴角左右摇摆，迷幻的青春，缤纷的色彩，放肆的创作，只有快乐是不变的主题。

Put all your burdens down, set yourself free and laugh a hearty laugh. Youth gleams with its illusions and no one can tell where it will go. So paint your life with funny colors and unbridled spirits. Only happiness is a constant theme.

女装 2015 春夏海派时尚流行趋势

多彩享梦
ENJOY COLORFUL DREAMS

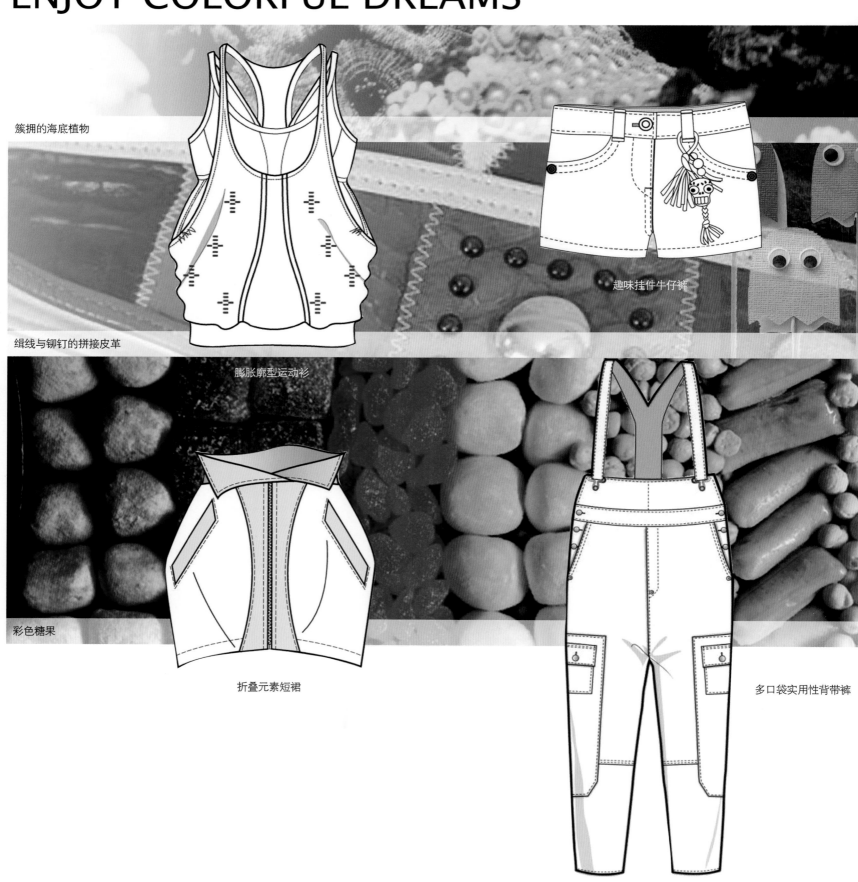

簇拥的海底植物

缉线与铆钉的拼接皮革

趣味挂件牛仔裤

膨胀廓型运动衫

彩色糖果

折叠元素短裙

多口袋实用性背带裤

多彩享梦
ENJOY COLORFUL DREAMS

时尚创意招贴

商城折扣广告

雨夜街景

个性涂鸦墙

扎染风格牛仔短裤

设计灵感

多彩与缤纷演绎出都市的节奏，霓虹灯雕饰出黑夜的娆美，空气中弥漫着梦想的味道。色彩的碰撞、风格的混搭，正如跳跃的音符行走在街头。

Colorful city beats, neon night beauty, the air is filled with the smell of dreams. Collision of colors and mix of styles on streets are like jumping notes on a score.

S女装

2015 春夏海派时尚流行趋势

多彩享梦
ENJOY COLORFUL DREAMS

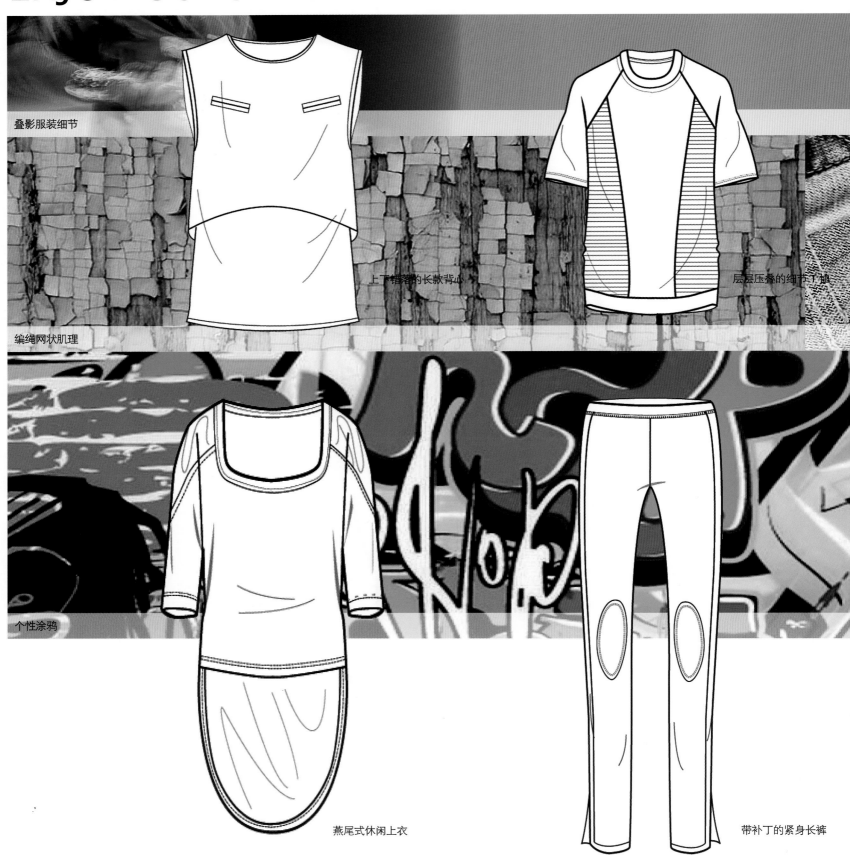

叠影服装细节

上下错落的长款背心

层层压叠的细节T恤

编绳网状肌理

个性涂鸦

燕尾式休闲上衣

带补丁的紧身长裤

2015 春夏海派时尚流行要素汇编·服装篇
2015 SPRING/SUMMER STYLE SHANGHAI FASHION ELEMENTS COLLECTION · APPAREL

海上星梦
STAR DREAMS OVER THE SEA

色彩

浪花白	青釉灰	青砖灰	乌梅黑	麦糖黄	胭脂红	枫叶红	紫檀棕
SH1200318	SH0801254	SH1101076	SH1000211	SH0101301	SH0303138	SH0303471	SH0303477

皮革黄	杏仁黄	琥珀橙	酸枝红	瓷瓦黑	雷雨灰	铁艺灰	青瓷灰
SH0101365	SH0101055	SH0200513	SH0303323	SH1100911	SH1100175	SH1100994	SH1100980

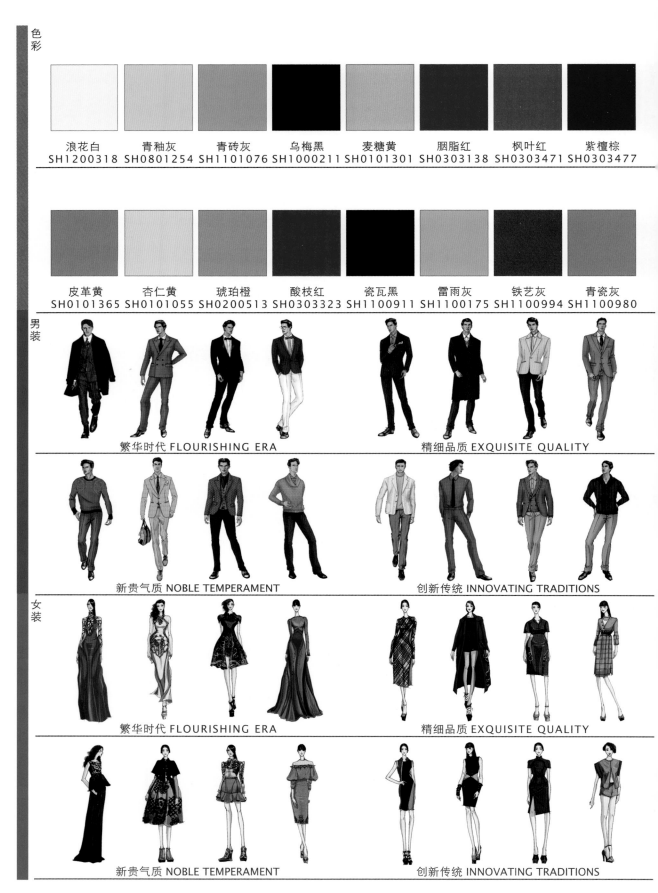

男装
繁华时代 FLOURISHING ERA　　精细品质 EXQUISITE QUALITY
新贵气质 NOBLE TEMPERAMENT　　创新传统 INNOVATING TRADITIONS

女装
繁华时代 FLOURISHING ERA　　精细品质 EXQUISITE QUALITY
新贵气质 NOBLE TEMPERAMENT　　创新传统 INNOVATING TRADITIONS

优活之梦
LIVING THE DREAM

色彩

| 孔雀蓝 | 青花蓝 | 豆浆白 | 水晶蓝 | 艾草绿 | 斗笠黄 | 香芋灰 | 紫砂棕 |
| SH0601053 | SH0503431 | SH0801123 | SH0502404 | SH0904102 | SH0802258 | SH0801388 | SH0701359 |

| 苔藓绿 | 蜜瓜黄 | 枣泥红 | 剑麻灰 | 抹茶绿 | 淡奶黄 | 蛋壳黄 | 曜石黑 |
| SH0903166 | SH0100139 | SH0301069 | SH0802256 | SH0901249 | SH0100230 | SH0201527 | SH1000027 |

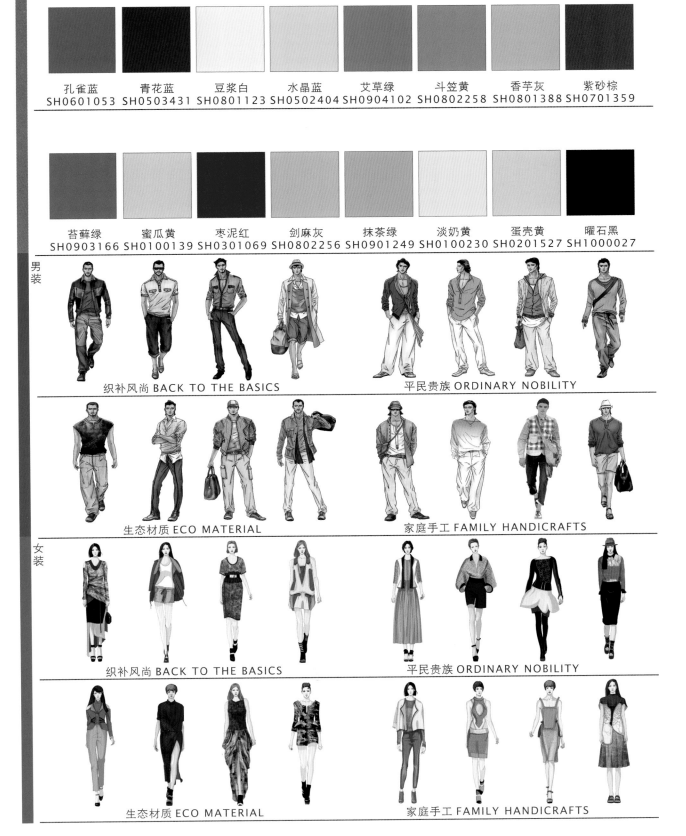

男装

织补风尚 BACK TO THE BASICS　　　平民贵族 ORDINARY NOBILITY

生态材质 ECO MATERIAL　　　家庭手工 FAMILY HANDICRAFTS

女装

织补风尚 BACK TO THE BASICS　　　平民贵族 ORDINARY NOBILITY

生态材质 ECO MATERIAL　　　家庭手工 FAMILY HANDICRAFTS

时空筑梦
REALISING DREAMS THROUGH SPACE

色彩

氯化灰	珍珠白	光影蓝	亚铜灰	柠檬黄	林荫绿	碳钢灰	天光蓝
SH0800262	SH0800210	SH0500285	SH0800481	SH0101109	SH0906065	SH1100763	SH0602025

日光白	陨石棕	恒星橙	烟云灰	苹果绿	沙砾黄	象牙黄	岩石灰
SH1200317	SH0701599	SH0201888	SH1100987	SH0900548	SH0800188	SH0801193	SH0800456

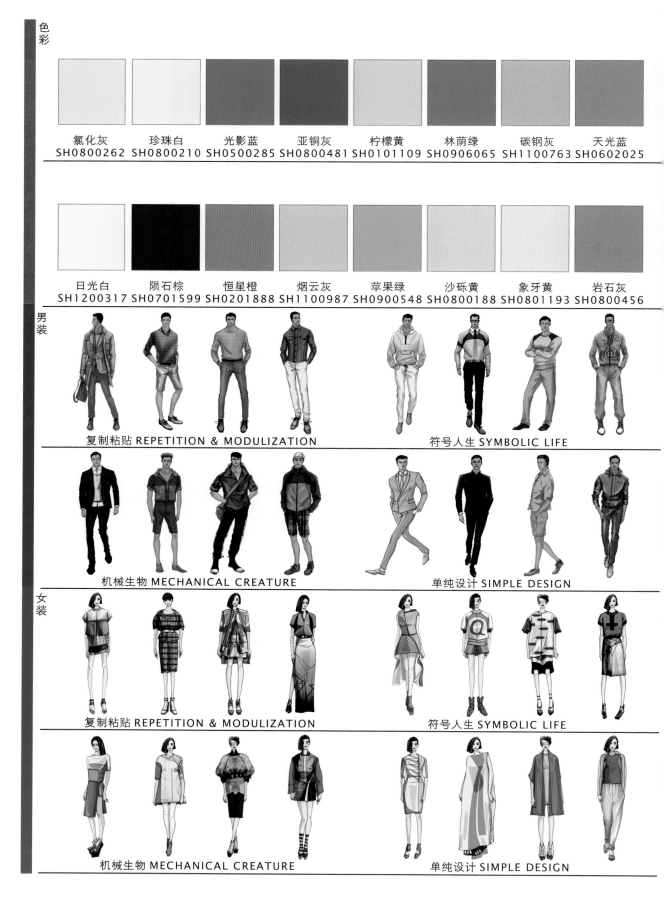

男装

复制粘贴 REPETITION & MODULIZATION　　符号人生 SYMBOLIC LIFE

机械生物 MECHANICAL CREATURE　　单纯设计 SIMPLE DESIGN

女装

复制粘贴 REPETITION & MODULIZATION　　符号人生 SYMBOLIC LIFE

机械生物 MECHANICAL CREATURE　　单纯设计 SIMPLE DESIGN

多彩享梦
ENJOY COLORFUL DREAMS

色彩

蜜桃粉	浅裸粉	幻彩红	深桃红	冰山蓝	泳池蓝	玛瑙蓝	松石绿
SH0402308	SH0801319	SH0300320	SH0402644	SH0602026	SH0503462	SH0600401	SH0601123

珊瑚红	迎春黄	琵琶橙	浅草绿	湖泊蓝	石榴红	霓虹绿	金币黄
SH0202768	SH0101222	SH0101118	SH0904042	SH0600419	SH0301062	SH0904632	SH0100081

男装

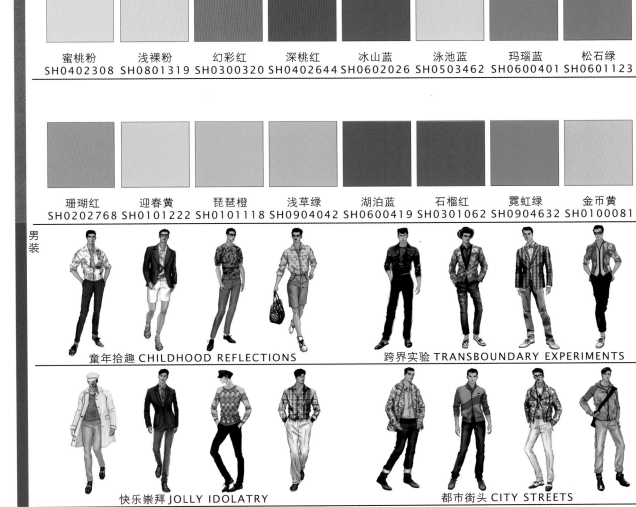

童年拾趣 CHILDHOOD REFLECTIONS　　跨界实验 TRANSBOUNDARY EXPERIMENTS

快乐崇拜 JOLLY IDOLATRY　　都市街头 CITY STREETS

女装

童年拾趣 CHILDHOOD REFLECTIONS　　跨界实验 TRANSBOUNDARY EXPERIMENTS

快乐崇拜 JOLLY IDOLATRY　　都市街头 CITY STREETS

图书在版编目（CIP）数据

海派时尚：2015春夏海派时尚流行趋势．服装篇／海派时尚流行趋势研究中心著．－－ 上海：东华大学出版社，2013.10
ISBN 978-7-5669-0382-2
Ⅰ．①海… Ⅱ．①海… Ⅲ．①服饰－市场预测－上海市－2015 Ⅳ．① TS941.13
中国版本图书馆CIP数据核字（2013）第 248244 号

责任编辑　马文娟
封面设计　范乃文　胡弘毅　周 吉

海派时尚：2015 春夏海派时尚流行趋势·服装篇
Haipai Shishang: 2015 Chunxia Haipai Shishang Liuxing Qushi·Fuzhuangpian
海派时尚流行趋势研究中心　著

出　　版：东华大学出版社（上海市延安西路1882号 邮政编码：200051）
出版社网址：http://www.dhupress.net
天猫旗舰店：http://dhdx.tmall.com
营销中心：021-62193056　62373056　62379558
印　　刷：上海中华商务联合印刷有限公司
开　　本：787mm×1092mm　1/8　印张：20
字　　数：526 千字
版　　次：2013 年 10 月第 1 版
印　　次：2013 年 10 月第 1 次印刷
书　　号：ISBN 978-7-5669-0382-2/TS·447
定　　价：580.00 元